# Yamaha RS/RXS 100 & 125 Singles Owners Workshop Manual

by Pete Shoemark
with an update Chapter covering 1977 on models
by Penelope A Cox

**Models covered**
RS100. 97cc. September 1974 to October 1983
RXS100. 98cc. April 1983 on
RS125. 123cc. September 1974 to January 1976
RS125 DX. 123cc. January 1976 to July 1984

**ISBN 1 85010 910 9**

© Haynes Publishing Group 1993

All rights reserved. No part of this book may be reproduced or transmitted in any form or by any means, electronic or mechanical, including photocopying, recording or by any information storage or retrieval system, without permission in writing from the copyright holder.

Printed in England (331 - 1S11)

**Haynes Publishing Group**
Sparkford Nr Yeovil
Somerset BA22 7JJ England

**Haynes Publications, Inc**
861 Lawrence Drive
Newbury Park
California 91320 USA

| British Library Cataloguing in Publication Data |
|---|
| A catalogue record for this book is available from the British Library |

| Library of Congress Catalog Card Number |
|---|
| 93-70494 |

# Acknowledgements

Our grateful thanks are due to R. S. Damerell and Son Ltd of St. Austell, who provided the service information necessary in the compilation of the manual, and to Fran Ridewood and Co, of Wells, who supplied the Yamaha RS100 model featured in the photographs.

Thanks are also due to Nelson James of Sparkford who provided the Yamaha RXS 100 featured on the front cover.

Martin Penny assisted with the stripdown and rebuilding and devised the ingenious methods for overcoming the lack of service tools.

Leon Martindale (Member of the Master Photographers Association) arranged and took the photographs which accompany the text, Les Brazier took the inside cover photographs.

Jeff Clew edited the text, and compiled the technical data used in the manual.

Finally, we would also like to thank the Avon Rubber Company, who kindly supplied information and technical assistance on tyre fitting; NGK Spark Plugs (UK) Ltd for information on spark plug maintenance and electrode conditions and Renold Limited for advice on chain care and renewal.

# About this manual

The author of this manual has the conviction that the only way in which a meaningful and easy to follow text can be written is first to do the work himself, under conditions similar to those found in the average household. As a result, the hands seen in the photographs are those of the author.

Unless specially mentioned, and therefore considered essential, special service tools have not been used. There is invariably some alternative means of loosening or removing a vital component when service tools are not available but risk of damage should always be avoided.

Each of the six Chapters is divided into numbered Sections. Within these Sections are numbered paragraphs. Cross reference throughout the manual is quite straightforward and logical. When reference is made 'See Section 6.10' it means Section 6, paragraph 10 in the same Chapter. If another Chapter is intended, the reference would read 'See Chapter 2, Section 6.10'.

All the photographs are captioned with a Section/paragraph number to which they refer, and are relevant to the Chapter text adjacent.

Figures (usually line illustrations) appear in a logical but numerical order, within a given Chapter. Fig. 1.1 therefore refers to the first figure in Chapter 1.

Left-hand and right-hand descriptions of the machines and their components refer to the left and right of a given machine when the rider is seated normally.

Motorcycle manufactures continually make changes to specifications and recommendations, and these, when notified, are incorporated into our manuals at the earliest opportunity.

We take great pride in the accuracy of information given in this manual, but motorcycle manufacturers make alterations and design changes during the production run of a particular motorcycle of which they do not inform us. No liability can be accepted by the authors or publishers for loss, damage or injury caused by any errors in, or omissions from, the information given.

# Introduction to the Yamaha RS 100/125 Singles

Although the history of Yamaha can be traced back to the year 1887, when a then very small company commenced manufacture of reed organs, it was not until 1954 that the company became interested in motor cycles. As can be imagined, the problems of marketing a motor cycle against a background of musical instruments manufacture were considerable. Some local racing successes and the use of hitherto unknown bright colour schemes helped achieve the desired results and in July 1955 the Yamaha Motor Company was established as a separate entity, employing a work force of less than 100 and turning out some 300 machines a month.

Competition successes continued and with the advent of tasteful styling that followed Italian trends, Yamaha became established as one of the world's leading motor cycle manufacturers. Part of this success story is the impressive list of Yamaha 'firsts' - a whole string of innovations that include electric starting, pressed steel frames, torque induction and 6 and 8 port engines. There is also the "Autolube" system of lubrication, in which the engine-driven oil pump is linked to the twist grip throttle, so that lubrication requirements are always in step with engine demands.

Since 1964, Yamaha has gained the World Championship on numerous occasions, in both the 125 cc and 250 cc classes. Indeed, Yamaha has dominated the lightweight classes in international road racing events to such an extent in recent years that several race promoters are now instituting a special type of event in their programme from which Yamaha machines are barred! Most of the racing successes have been achieved with twin cylinder two-strokes and the practical experience gained has been applied to the road going versions.

The RS100 and 125 singles were introduced to fulfill a need for an economical lightweight motorcycle for the commuter. The compact design makes for a nimble machine in traffic, the small engine providing surprisingly brisk performance due to the use of a five-speed gearbox and the torque induction system. In addition to the basic drum-braked models, there is a DX model, the specifications of which include a tachometer and an hydraulic disc front brake.

For information relating to models introduced from 1977 on refer to Chapter 7.

# Contents

| | Page |
|---|---|
| Acknowledgements | 2 |
| About this manual | 2 |
| Introduction to the Yamaha RS100 and 125 Singles | 2 |
| Safety first! | 5 |
| Ordering spare parts | 6 |
| Routine maintenance | 7 |
| Quick glance maintenance, adjustments and capacities | 11 |
| Working conditions and tools | 12 |
| Recommended torque settings and lubricants | 13 |
| Chapter 1 Engine, clutch and gearbox | 14 |
| Chapter 2 Fuel system and lubrication | 44 |
| Chapter 3 Ignition system | 56 |
| Chapter 4 Frame and forks | 61 |
| Chapter 5 Wheels, brakes and tyres | 76 |
| Chapter 6 Electrical system | 93 |
| Chapter 7 The 1977 on models | 102 |
| Wiring diagrams | 114 |
| Conversion factors | 120 |
| Index | 121 |

1977 Yamaha RS100 model

1976 Yamaha RS 125 model

# Safety first!

Professional motor mechanics are trained in safe working procedures. However enthusiastic you may be about getting on with the job in hand, do take the time to ensure that your safety is not put at risk. A moment's lack of attention can result in an accident, as can failure to observe certain elementary precautions.

There will always be new ways of having accidents, and the following points do not pretend to be a comprehensive list of all dangers; they are intended rather to make you aware of the risks and to encourage a safety-conscious approach to all work you carry out on your vehicle.

## Essential DOs and DON'Ts

**DON'T** start the engine without first ascertaining that the transmission is in neutral.
**DON'T** suddenly remove the filler cap from a hot cooling system – cover it with a cloth and release the pressure gradually first, or you may get scalded by escaping coolant.
**DON'T** attempt to drain oil until you are sure it has cooled sufficiently to avoid scalding you.
**DON'T** grasp any part of the engine, exhaust or silencer without first ascertaining that it is sufficiently cool to avoid burning you.
**DON'T** allow brake fluid or antifreeze to contact the machine's paintwork or plastic components.
**DON'T** syphon toxic liquids such as fuel, brake fluid or antifreeze by mouth, or allow them to remain on your skin.
**DON'T** inhale dust – it may be injurious to health (see *Asbestos* heading).
**DON'T** allow any spilt oil or grease to remain on the floor – wipe it up straight away, before someone slips on it.
**DON'T** use ill-fitting spanners or other tools which may slip and cause injury.
**DON'T** attempt to lift a heavy component which may be beyond your capability – get assistance.
**DON'T** rush to finish a job, or take unverified short cuts.
**DON'T** allow children or animals in or around an unattended vehicle.
**DON'T** inflate a tyre to a pressure above the recommended maximum. Apart from overstressing the carcase and wheel rim, in extreme cases the tyre may blow off forcibly.
**DO** ensure that the machine is supported securely at all times. This is especially important when the machine is blocked up to aid wheel or fork removal.
**DO** take care when attempting to slacken a stubborn nut or bolt. It is generally better to pull on a spanner, rather than push, so that if slippage occurs you fall away from the machine rather than on to it.
**DO** wear eye protection when using power tools such as drill, sander, bench grinder etc.
**DO** use a barrier cream on your hands prior to undertaking dirty jobs – it will protect your skin from infection as well as making the dirt easier to remove afterwards; but make sure your hands aren't left slippery. Note that long-term contact with used engine oil can be a health hazard.
**DO** keep loose clothing (cuffs, tie etc) and long hair well out of the way of moving mechanical parts.
**DO** remove rings, wristwatch etc, before working on the vehicle – especially the electrical system.
**DO** keep your work area tidy – it is only too easy to fall over articles left lying around.
**DO** exercise caution when compressing springs for removal or installation. Ensure that the tension is applied and released in a controlled manner, using suitable tools which preclude the possibility of the spring escaping violently.
**DO** ensure that any lifting tackle used has a safe working load rating adequate for the job.
**DO** get someone to check periodically that all is well, when working alone on the vehicle.
**DO** carry out work in a logical sequence and check that everything is correctly assembled and tightened afterwards.
**DO** remember that your vehicle's safety affects that of yourself and others. If in doubt on any point, get specialist advice.
**IF,** in spite of following these precautions, you are unfortunate enough to injure yourself, seek medical attention as soon as possible.

## Asbestos

Certain friction, insulating, sealing, and other products – such as brake linings, clutch linings, gaskets, etc – contain asbestos. *Extreme care must be taken to avoid inhalation of dust from such products since it is hazardous to health.* If in doubt, assume that they *do* contain asbestos.

## Fire

Remember at all times that petrol (gasoline) is highly flammable. Never smoke, or have any kind of naked flame around, when working on the vehicle. But the risk does not end there – a spark caused by an electrical short-circuit, by two metal surfaces contacting each other, by careless use of tools, or even by static electricity built up in your body under certain conditions, can ignite petrol vapour, which in a confined space is highly explosive.

Always disconnect the battery earth (ground) terminal before working on any part of the fuel or electrical system, and never risk spilling fuel on to a hot engine or exhaust.

It is recommended that a fire extinguisher of a type suitable for fuel and electrical fires is kept handy in the garage or workplace at all times. Never try to extinguish a fuel or electrical fire with water.

**Note:** *Any reference to a 'torch' appearing in this manual should always be taken to mean a hand-held battery-operated electric lamp or flashlight. It does **not** mean a welding/gas torch or blowlamp.*

## Fumes

Certain fumes are highly toxic and can quickly cause unconsciousness and even death if inhaled to any extent. Petrol (gasoline) vapour comes into this category, as do the vapours from certain solvents such as trichloroethylene. Any draining or pouring of such volatile fluids should be done in a well ventilated area.

When using cleaning fluids and solvents, read the instructions carefully. Never use materials from unmarked containers – they may give off poisonous vapours.

Never run the engine of a motor vehicle in an enclosed space such as a garage. Exhaust fumes contain carbon monoxide which is extremely poisonous; if you need to run the engine, always do so in the open air or at least have the rear of the vehicle outside the workplace.

## The battery

Never cause a spark, or allow a naked light, near the vehicle's battery. It will normally be giving off a certain amount of hydrogen gas, which is highly explosive.

Always disconnect the battery earth (ground) terminal before working on the fuel or electrical systems.

If possible, loosen the filler plugs or cover when charging the battery from an external source. Do not charge at an excessive rate or the battery may burst.

Take care when topping up and when carrying the battery. The acid electrolyte, even when diluted, is very corrosive and should not be allowed to contact the eyes or skin.

If you ever need to prepare electrolyte yourself, always add the acid slowly to the water, and never the other way round. Protect against splashes by wearing rubber gloves and goggles.

## Mains electricity and electrical equipment

When using an electric power tool, inspection light etc, always ensure that the appliance is correctly connected to its plug and that, where necessary, it is properly earthed (grounded). Do not use such appliances in damp conditions and, again, beware of creating a spark or applying excessive heat in the vicinity of fuel or fuel vapour. Also ensure that the appliances meet the relevant national safety standards.

## Ignition HT voltage

A severe electric shock can result from touching certain parts of the ignition system, such as the HT leads, when the engine is running or being cranked, particularly if components are damp or the insulation is defective. Where an electronic ignition system is fitted, the HT voltage is much higher and could prove fatal.

# Ordering spare parts

When ordering spare parts for the Yamaha RS models, it is advisable to deal direct with an official Yamaha agent, who will be able to supply many of the items required ex-stock. Although parts can be ordered from Yamaha direct, it is preferable to route the order via a local agent even if the parts are not available from stock. He is in a better position to specify exactly the parts required and to identify the relevant spare part numbers so that there is less chance of the wrong part being supplied by the manufacturer due to a vague or incomplete description.

When ordering spares, always quote the frame and engine numbers in full, together with any prefixes or suffixes in the form of letters. The frame number is found stamped on the right-hand side of the steering head, in line with the forks. The engine number is stamped on the left-hand side of the upper crankcase, immediately below the left-hand carburettor.

Use only parts of genuine Yamaha manufacture. A few pattern parts are available, sometimes at cheaper prices, but there is no guarantee that they will give such good service as the originals they replace. Retain any worn or broken parts until the replacements have been obtained; they are sometimes needed as a pattern to help identify the correct replacement when design changes have been made during a production run.

Some of the more expendable parts such as spark plugs, bulbs, tyres, oils and greases etc., can be obtained from accessory shops and motor factors, who have convenient opening hours, charge lower prices and can often be found not far from home. It is also possible to obtain parts on a Mail Order basis from a number of specialists who advertise regularly in the motor cycle magazines.

**Frame number location**

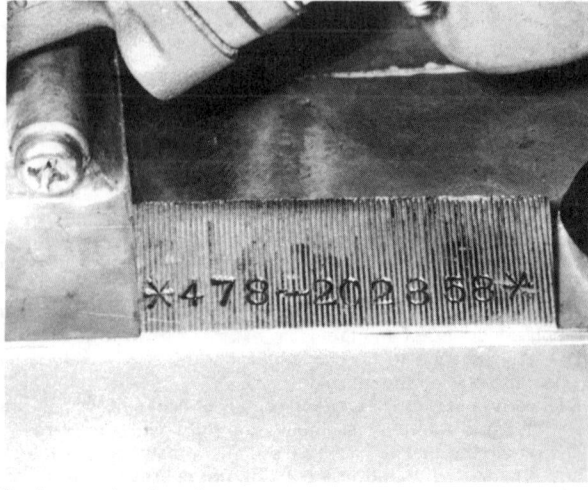

Engine number location

# Routine maintenance

*For information relating to 1977 on models refer to Chapter 7*

## Introduction

Periodic routine maintenance is a continuous process that commences immediately the machine is used. It must be carried out at specified mileage recording, or on a calendar basis if the machine is not used frequently, whichever is the sooner. Maintenance should be regarded as an insurance policy, to help keep the machine in the peak of condition and to ensure long, trouble-free service. It has the additional benefit of giving early warning of any faults that may develop and will act as a regular safety check, to the obvious advantage of both rider and machine alike.

The various maintenance tasks are described under their respective mileage and calendar headings. Accompanying diagrams are provided, where necessary. It should be remembered that the interval between the various maintenance tasks serves only as a guide. As the machine gets older or is used under particularly adverse conditions, it would be advisable to reduce the period between each check.

For ease of reference each service operation is described in detail under the relevant heading. However, if further general information is required, it can be found within the manual under the pertinent section heading in the relevant Chapter.

In order that the routine maintenance tasks are carried out with as much ease as possible, it is essential that a good selection of general workshop tools are available.

Included in the kit must be a range of metric ring or combination spanners, a selection of crosshead screwdrivers and at least one pair of circlip pliers.

Additionally, owing to the extreme tightness of most casing screws on Japanese machines, an impact screwdriver, together with a choice of large or small crosshead screw bits, is absolutely indispensable. This is particularly so if the engine has not been dismantled since leaving the factory.

## Weekly or every 200 miles

### 1 Topping up engine oil

Check that there is sufficient lubricant in the oil tank which feeds the mechanical oil pump. All models are fitted with an oil level sight glass, below the level of which the oil must not be allowed to fall.

The oil filler is located under the dualseat on all models. Fill to within about 25 mm (one inch) of the filler neck, with SAE 30 two-stroke oil.

### 2 Tyre pressures

Check the tyre pressures with a pressure gauge that is known to be accurate. Always check the pressure when the tyres are cold. If the machine has travelled a number of miles, the tyres will have become hotter and consequently the pressure will have increased. A false reading will therefore result.

Fill oil tank with two-stroke oil

| Tyre pressures | Solo | With pillion passenger or high speed touring |
|---|---|---|
| Front | 1.5 kg/cm$^2$ (21 psi) | 1.8 kg/cm$^2$ (26 psi) |
| Rear | 2.0 kgs/cm$^2$ (28 psi) | 2.3 kgs/cm$^2$ (33 psi) |

### 3 Hydraulic fluid level - disc brake models

Check the level of the fluid in the master cylinder reservoir. On machines fitted with a tamper-proof cover, this may be observed through the translucent side of the reservoir. On models fitted with a screw cap, remove the cap to observe the fluid level. During normal service, it is unlikely that the hydraulic fluid level will fall dramatically, unless a leak has developed in the system. If this occurs, the fault should be remedied **at once**. The level will fall slowly as the brake linings wear and the fluid deficiency should be corrected, when required. Always use an hydraulic fluid of DOT 3 or SAE J1703 specification, and do not mix different types of fluid, even if the specifications appear the same. This will preclude the possibility of two incompatible fluids being mixed and the resultant chemical reaction damaging the seals.

If the level in the reservoir has been allowed to fall below the specified limit, and air has entered the system, the brake in question must be bled, as described in Chapter 5, Section 10.

### 4 Transmission oil

Unscrew the filler cap on the right-hand engine cover and check the transmission oil level by means of the integral dipstick. Replenish, if necessary, with SAE 10W/30 engine oil.

## Routine maintenance

### 5 Battery electrolyte level

1 An FB or a GS battery is fitted as standard. This battery is a lead-acid type and has a capacity of 4 amp hours.

2 The transparent plastic case of the battery permits the upper and lower levels of the electrolyte to be observed when the battery is lifted from its housing below the dualseat. Maintenance is normally limited to keeping the electrolyte level between the prescribed upper and lower limits and by making sure that the vent pipe is not blocked. The lead plates and their separators can be seen through the transparent case, a further guide to the general condition of the battery.

3 Unless acid is spilt, as may occur if the machine falls over, the electrolyte should always be topped up with distilled water, to restore the correct level. If acid is spilt on any part of the machine, it should be neutralised with an alkali such as washing soda and washed away with plenty of water, otherwise serious corrosion will occur. Top up with sulphuric acid of the correct specific gravity (1.260 - 1.280) only when spillage has occurred. Check that the vent pipe is well clear of the frame tubes or any of the other cycle parts, for obvious reasons.

### 6 Control cable lubrication

Apply a few drops of motor oil to the exposed inner portion of each control cable. This will prevent drying-up of the cables between the more thorough lubrication that should be carried out during the 2000 mile/3 monthly service.

### 7 Rear chain lubrication and adjustment

In order that the life of the rear chain be extended as much as possible, regular lubrication and adjustment is essential.

Intermediate lubrication should take place at the weekly or 200 mile service interval with the chain in situ. Application of one of the proprietary chain greases contained in an aerosol can is ideal. Ordinary engine oil can be used, though owing to the speed with which it is flung off the rotating chain, its effectiveness is limited.

Adjust the chain after lubrication, so that there is approximately 20 mm (¾ in) slack in the middle of the lower run. Always check with the chain at the tightest point as a chain rarely wears evenly during service.

Adjustment is accomplished after placing the machine on the centre stand and slackening the wheel nut, so that the wheel can be drawn backwards by means of the drawbolt adjusters in the fork ends.

The torque arm nuts and the rear brake adjuster must also be slackened during this operation. Adjust the drawbolts an equal amount to preserve wheel alignment. The fork ends are clearly marked with a series of parallel lines above the adjusters, to provide a simple visual check.

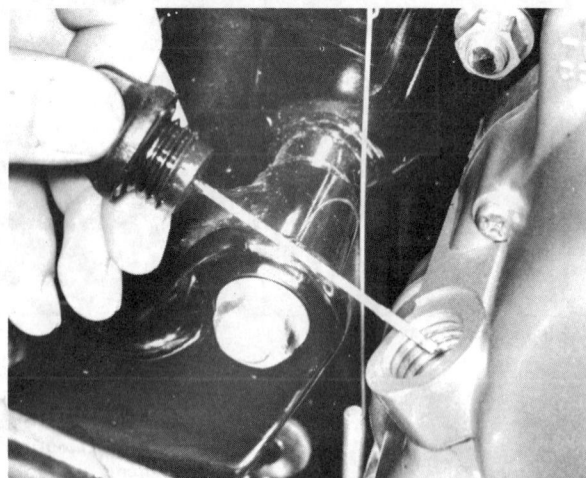

Gearbox oil level can be checked against dipstick

Battery is located beneath dualseat

Use aerosol chain lubricant frequently

Rear chain is adjusted by drawbolts

## Routine maintenance

### 8 Safety check
Give the machine a close visual inspection, checking for loose nuts and fittings, frayed control cables etc. Check the tyres for damage, especially splitting on the sidewalls. Remove any stones or other objects caught between the treads. This is particularly important on the front tyre, where rapid deflation due to penetration of the inner tube will almost certainly cause total loss of control.

### 9 Legal check
Ensure that the lights, horn and trafficators function correctly, also the speedometer.

### 10 Brake check
Check that both brakes function effectively and operate smoothly. Check the operation of the stop lamp switches.

On disc brake models inspect the system for leakage and check the fluid level as described previously in this section. On drum brakes, check that the operating mechanism is correctly adjusted, see Chapter 5.

### Three monthly or every 2000 miles

Carry out the checks listed under the weekly/200 mile heading and then complete the following:

#### 1 Checking the contact breaker gap
Rotate the engine until the contact breakers are in the fully open position. The correct gap is within the range 0.3 - 0.4 mm (0.012 - 0.016 in). Adjustment is effected by slackening the screw holding the fixed contact breaker point in position and moving the point either closer to or further away with a screwdriver inserted between the small upright post and the slot in the fixed contact plate. Make sure that the points are in the fully open position when this adjustment is made or a false reading will result. When the gap is correct, tighten the screw and recheck.

#### 2 Checking and resetting the ignition timing
If the ignition timing is correct, the contact breaker points must be on the verge of separation when the piston is between 1.3 and 2.0 mm (0.0512 - 0.0787 in) before top dead centre.

The contact breaker can be adjusted within reason to obtain separation at this point, but it should be noted that if badly worn, it will have to be renewed in order to achieve correct timing and gap clearance. Reference should be made to Chapter 3 for further details.

#### 3 Cleaning the air filter
A composite fabric/foam air cleaner element is located within a sealed box on all models. It is essential that this is kept clean and properly lubricated to ensure efficient operation. It is located behind the right-hand side panel on all models, and is retained by a single screw.

Remove the element from the case, and clean it thoroughly by washing it in petrol. Dry carefully by squeezing it in a piece of clean rag, ensuring that all dirt and residual petrol is removed.

Soak the element in clean engine oil, working it into the foam. Squeeze out any excess oil so that the element is damp, but not dripping.

Slide the relubricated element onto its wire mesh or plastic support frame, and replace it in the air cleaner case. Refit the cover and side panel.

Do not on any account run the machine with the air filter removed or with the air cleaner hoses disconnected. If this precaution is not observed, the engine will run with a permanently weak mixture, which will cause overheating and possible seizure.

#### 4 Cleaning the exhaust system
Due to the oily nature of the exhaust gases of a two-stroke engine, the exhaust system will progressively block up with sludge and hard carbon deposits, as the machine is used. As the

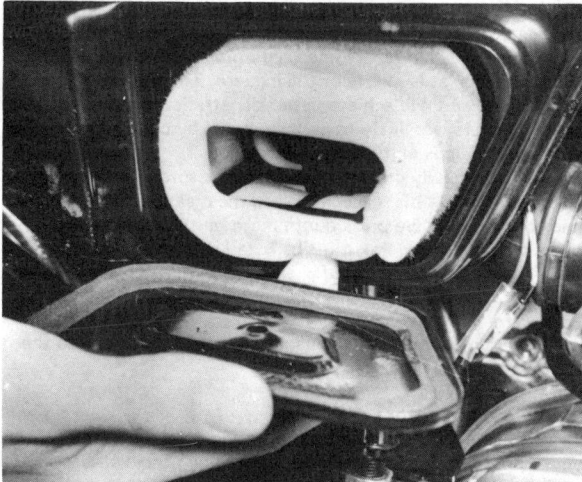

Remove air filter element for cleaning

sludge builds up, back pressure will increase, with a resulting fall-off in performance. The deposits must therefore be removed at regular intervals.

To aid cleaning, the silencer is fitted with a detachable baffle, which is retained by a single screw, at the rear of the silencer. After removal of the screw, the baffle may be withdrawn by applying a stout pair of pliers to the crossbar in the baffle end, and twisting the baffle as necessary.

If the build-up of carbon and oil is not too great, a wash with a petrol/paraffin mixture will probably suffice as the cleaning medium. Otherwise, more drastic action will be necessary, such as the application of a blowlamp flame to burn away the accumulated deposits. After a number of cleaning operations, the fibreglass or asbestos wool with which the rearmost section of the baffle is shrouded, may disintegrate. Replacement of the wool is not strictly necessary as it will not affect performance but a slight increase in exhaust noise may be experienced.

When refitting the baffle, ensure that the retaining screw is tightened fully. If the screw falls out, the baffle will follow, creating excessive noise accompanied by a marked fall-off in performance. Running the engine without the baffle will also give rise to a weak mixture, causing overheating and possible engine seizure.

#### 5 Oil pump adjustment
Remove the cover and three retaining screws at the front of the right-hand outer cover, exposing the oil pump. The throttle valve should be arranged so that it is just closed, having first set the throttle cable free play to 0.5 – 1.0 mm.

With the throttle in this position, the index mark on the outer face of the oil pump pulley should be exactly in line with the roll pin protruding from the pump plunger. If the alignment is not correct, loosen the locknut on the oil pump control cable adjuster and rotate the adjuster, as required, to bring the two marks into line. Tighten the locknut, open and shut the throttle a number of times and recheck the alignment.

#### 6 Final drive chain lubrication
The final drive chain should be removed from the machine for thorough cleaning and lubrication if long service life is to be expected. This is in addition to the intermediate lubrication carried out with the chain on the machine, as described under the weekly/200 mile service heading.

Separate the chain by removing the master link and run it off the sprockets. If an old chain is available, interconnect the old and new chain, before the new chain is run off the sprockets. In this way the old chain can be pulled into place on the sprockets and then used to pull the regreased chain into place with ease.

# Routine maintenance

Clean the chain thoroughly in a paraffin bath and then finally with petrol. The petrol will wash the paraffin out of the links and rollers which will then dry more quickly.

Allow the chain to dry and then immerse it in a molten lubricant such as Linklyfe or Chainguard. These lubricants must be used hot and will achieve better penetration of the links and rollers and are less likely to be thrown off by centrifugal force when the chain is in motion.

Refit the newly greased chain onto the sprocket, replacing the master link. This is accomplished most easily when the free ends of the chain are pushed into mesh on the rear wheel sprocket. The spring link must be fitted so that the closed end faces the normal direction of chain travel.

### 7 General lubrication

Apply grease or oil to the handlebar lever pivots and to the centre stand and prop stand pivots.

### 8 Control cable lubrication

Lubricate the control cables thoroughly with motor oil or an all-purpose oil. A good method of lubricating the cables is shown in the accompanying illustration, using a plasticine funnel. This method has the disadvantage that the cables usually need removing from the machine. An hydraulic cable oiler which pressurises the lubricant overcomes this problem. Do not lubricate nylon lined cables (which may have been fitted as replacements), as the oil may cause the nylon to swell, thereby causing total cable seizure.

### 10 Changing the transmission oil

Place a container of greater than 800 cc capacity below the gearbox and remove the oil filler plug from the right-hand engine cover. Unscrew the gearbox drain plug and allow the lubricant to drain. The oil should be drained with the engine hot, ideally after the machine has been on a run, as the lubricant will be thinner and so drain more rapidly and completely.

When drainage is complete, refit the drain plug and the sealing washer and refill the gearbox with approximately 700 - 750 cc of SAE 10W/30 engine oil. Allow the oil to settle and then check the level with the filler plug integral dipstick. When checking the level, do not screw the filler plug home, but allow it to rest on the orifice edge.

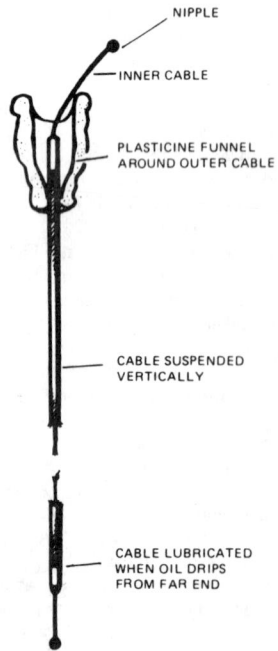

**Control cable oiling**

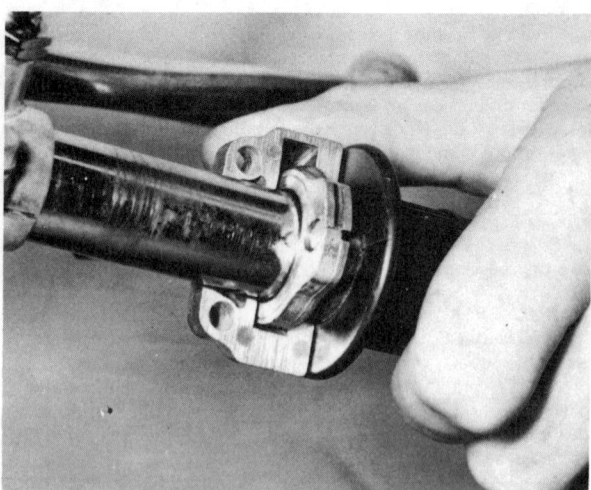

Lubricate cables and controls

---

### Six monthly or every 4000 miles

Complete the checks listed under the weekly/200 mile and three monthly/2000 mile headings, then complete the following additional procedures.

### 1 Changing the front fork damping oil

Place the machine on the centre stand so that the front wheel is clear of the ground. Place wooden blocks below the crankcase in order to prevent the machine from tipping forward. Loosen and remove the chrome cap bolts. Unscrew the drain plug from each fork leg, located directly above the wheel spindle, and allow the damping fluid to drain into a suitable container. This is accomplished most easily if the legs are attended to in turn. Take care not to spill any fluid onto the brake disc or tyre. The forks may be pumped up and down slowly to expel any remaining fluid. Refit and tighten the drain plugs. Refill each fork leg with 150 cc of good quality fork oil. Refit and tighten the chrome cap bolts and then tighten the two pinch bolts.

### 2 Wheel condition

Check the spoke tension by gently tapping each one with a metal object. A loose spoke is identifiable by the low pitch noise generated. If any spoke needs considerable tightening, it will be necessary to remove the tyre and inner tube in order to file down the protruding spoke end. This will prevent the spoke from chafing through the rim band and piercing the inner tube. Rotate the wheel and test for rim runout. Excessive runout will cause handling problems and should be corrected by tightening or loosening the relevant spokes. Care must be taken, since altering the tension in the wrong spokes may create more problems.

### 3 Cleaning the fuel tap

Because cleaning of the fuel tap requires that the tank be drained first, this operation should be carried out, if possible when the fuel is at a low ebb, NOT at high tide! Disconnect the fuel feed pipe at the carburettor union after releasing the tension of the spring clip. Drain the fuel into a clean container, with the tap in the 'reserve' position.

Remove the filter (sediment trap) bowl from the main body of the fuel tap by applying a suitable spanner to the hexagon. Note the 'O' ring. The tap body screws into the underside of the tank. Clean the filter bowl and the filter matrix in clean fuel and then replace the components. If the condition of the 'O' ring or the tap/tank rubber seal is suspect, it should be renewed.

## Routine maintenance

*4 Decarbonising the cylinder head, barrel and piston.*
Removal of the cylinder components, decarbonising and inspection, should be carried out by referring to the relevant Sections in Chapter 1. This work can be accomplished without removing the engine from the frame.

*5 Removal, inspection and relubrication of wheel bearings*
Carry out the operations listed in the heading by following the procedure given in Chapter 5, Section 13 for front wheel and Section 14 for the rear wheel.

### General maintenance adjustments

*1 Clutch adjustment*
The intervals at which the clutch should be adjusted will depend on the style of riding and the conditions under which the machine is used.
Adjust the clutch in two stages as follows:
Remove the left-hand outer cover. Loosen the cable adjuster screw locknut and turn the adjuster inwards fully, to give plenty of slack in the inner cable. Loosen the adjuster screw locknut in the casing and turn the screw clockwise until slight resistance is felt. Back off the screw about ¼ turn and tighten the locknut. The cover may be replaced.
Undo the cable adjuster screw at the handlebar lever, until there is approximately 2 - 3 mm (0.08 - 0.12 in) play measured between the inner face of the lever and the stock face. Finally, tighten the cable adjuster locknut.

*2 Checking brake pad wear (disc brake models)*
Brake pad wear depends largely on the conditions in which the machine is ridden and at what speed. It is difficult therefore, to give precise inspection intervals, but it follows that pad wear should be checked more frequently on a hard ridden machine.
The inner (fixed) pad is located by a screw and support plate, which should be removed to release it. The outer (moving) pad can then be eased out, using a small screwdriver or similar tool. It will be noted that the inner pad can easily be identified by the small locating peg, which engages in a slot in the caliper, and also the hole for the retaining screw. Examine the pads for wear:

| Pad type | Nominal size | Wear limit |
|---|---|---|
| Inner (fixed) | 12.0 mm | 7.5 mm |
| Outer (moving) | 15.5 mm | 11.0 mm |

Should the pad be found to be worn below the limits given, it should be renewed. Look also for signs of staining on the friction material. This may be caused by leakage from the fork

Clutch pushrod adjuster is held by locknut

leg or from the caliper seals; in either case attention must be given to locating and rectifying the source of the leak. Ensure that the piston is not inadvertently expelled whilst the pads are removed. Should this occur, air will enter the system and will have to be bled out. Reassembly is a direct reversal of the removal sequence. Be sure that the pads, if re-used, are kept clean and that no foreign matter is allowed to enter the caliper. Ensure that the fixed pad locates in its groove correctly and that the brake is functioning properly, before using the machine on the road.

*3 Checking brake lining wear, all models*
The rear drum brake linings, and the front drum brake linings, where fitted, should be checked for wear after detaching the relevant wheel as described in Chapter 5, Section 4 or 17. The amount of wear can be assessed by measuring the diameter of the shoes whilst in place on the brake plate, and comparing the reading with the following nominal sizes:

|  | RS100 | RS125 |
|---|---|---|
| Front brake shoe diameter: | 110 mm | 150 mm |
| Rear brake shoe diameter: | 110 mm | 130 mm |

If the measured diameter is 5 mm, or more, smaller, this indicates the need for renewal.

# Quick glance:
# Maintenance, adjustments and capacities

| | |
|---|---|
| **Engine (oil tank)** | SAE 30 two-stroke oil. Separate lubrication system operated by 'Autolube' oil pump, interconnected with the throttle |
| | Capacity 1.5 litres (1.6 US qts) (2.6 Imp pints) or 1.3 litres (1.4 US qts) (2.3 Imp pints) |
| **Gearbox** | Capacity 750 cc (0.8 US qts) (1.3 Imp pints) |
| **Front forks** | 150 cc per fork leg fork oil (0.16 US qts) (0.26 Imp pints) |
| **Contact breaker gap** | 0.3 to 0.4 mm (0.012 to 0.016 in) |
| **Spark plug gap** | 0.5 - 0.6 mm (0.020 - 0.024 in) |
| **Tyre pressures:** | Front / Rear |
| Solo | 1.5 kgs/cm$^2$ (21 psi) / 2.0 kgs/cm$^2$ (28 psi) |
| Pillion, or high speed | 1.8 kgs/cm$^2$ (26 psi) / 2.3 kgs/cm$^2$ (33 psi) |

# Working conditions and tools

When a major overhaul is contemplated, it is important that a clean, well-lit working space is available, equipped with a workbench and vice, and with space for laying out or storing the dismantled assemblies in an orderly manner where they are unlikely to be disturbed. The use of a good workshop will give the satisfaction of work done in comfort and without haste, where there is little chance of the machine being dismantled and reassembled in anything other than clean surroundings. Unfortunately, these ideal working conditions are not always practicable and under these latter circumstances when improvisation is called for, extra care and time will be needed.

The other essential requirement is a comprehensive set of good quality tools. Quality is of prime importance since cheap tools will prove expensive in the long run if they slip or break when in use, causing personal injury or expensive damage to the component being worked on. A good quality tool will last a long time, and more than justify the cost.

For practically all tools, a tool factor is the best source since he will have a very comprehensive range compared with the average garage or accessory shop. Having said that, accessory shops often offer excellent quality tools at discount prices, so it pays to shop around. There are plenty of tools around at reasonable prices, but always aim to purchase items which meet the relevant national safety standards. If in doubt, seek the advice of the shop proprietor or manager before making a purchase.

The basis of any tool kit is a set of open-ended spanners, which can be used on almost any part of the machine to which there is reasonable access. A set of ring spanners makes a useful addition, since they can be used on nuts that are very tight or where access is restricted. Where the cost has to be kept within reasonable bounds, a compromise can be effected with a set of combination spanners – open-ended at one end and having a ring of the same size on the other end. Socket spanners may also be considered a good investment, a basic 3/8 in or 1/2 in drive kit comprising a ratchet handle and a small number of socket heads, if money is limited. Additional sockets can be purchased, as and when they are required. Provided they are slim in profile, sockets will reach nuts or bolts that are deeply recessed. When purchasing spanners of any kind, make sure the correct size standard is purchased. Almost all machines manufactured outside the UK and the USA have metric nuts and bolts, whilst those produced in Britain have BSF or BSW sizes. The standard used in USA is AF, which is also found on some of the later British machines. Others tools that should be included in the kit are a range of crosshead screwdrivers, a pair of pliers and a hammer.

When considering the purchase of tools, it should be remembered that by carrying out the work oneself, a large proportion of the normal repair cost, made up by labour charges, will be saved. The economy made on even a minor overhaul will go a long way towards the improvement of a toolkit.

In addition to the basic tool kit, certain additional tools can prove invaluable when they are close to hand, to help speed up a multitude of repetitive jobs. For example, an impact screwdriver will ease the removal of screws that have been tightened by a similar tool, during assembly, without a risk of damaging the screw heads. And, of course, it can be used again to retighten the screws, to ensure an oil or airtight seal results. Circlip pliers have their uses too, since gear pinions, shafts and similar components are frequently retained by circlips that are not too easily displaced by a screwdriver. There are two types of circlip pliers, one for internal and one for external circlips. They may also have straight or right-angled jaws.

One of the most useful of all tools is the torque wrench, a form of spanner that can be adjusted to slip when a measured amount of force is applied to any bolt or nut. Torque wrench settings are given in almost every modern workshop or service manual, where the extent to which a complex component, such as a cylinder head, can be tightened without fear of distortion or leakage. The tightening of bearing caps is yet another example. Overtightening will stretch or even break bolts, necessitating extra work to extract the broken portions.

As may be expected, the more sophisticated the machine, the greater is the number of tools likely to be required if it is to be kept in first class condition by the home mechanic. Unfortunately there are certain jobs which cannot be accomplished successfully without the correct equipment and although there is invariably a specialist who will undertake the work for a fee, the home mechanic will have to dig more deeply in his pocket for the purchase of similar equipment if he does not wish to employ the services of others. Here a word of caution is necessary, since some of these jobs are best left to the expert. Although an electrical multimeter of the AVO type will prove helpful in tracing electrical faults, in inexperienced hands it may irrevocably damage some of the electrical components if a test current is passed through them in the wrong direction. This can apply to the synchronisation of twin or multiple carburettors too, where a certain amount of expertise is needed when setting them up with vacuum gauges. These are, however, exceptions. Some instruments, such as a strobe lamp, are virtually essential when checking the timing of a machine powered by CDI ignition system. In short, do not purchase any of these special items unless you have the experience to use them correctly.

Although this manual shows how components can be removed and replaced without the use of special service tools (unless absolutely essential), it is worthwhile giving consideration to the purchase of the more commonly used tools if the machine is regarded as a long term purchase Whilst the alternative methods suggested will remove and replace parts without risk of damage, the use of the special tools recommended and sold by the manufacturer will invariably save time.

# Recommended torque settings:

| | |
|---|---|
| Cylinder head nuts | 1.8 to 2.5 m - kgs (13 to 18 ft lbs) |
| Generator flywheel nut | 4.0 to 4.5 m - kgs (29 to 32.5 ft lbs) |
| Clutch centre nut | 3.0 to 5.0 m - kgs (21.7 to 36 ft lbs) |
| Gearbox sprocket nut | 6.4 to 10.0 m - kgs (46 to 72 ft lbs) |
| Crankcase screws | 1.1 to 1.3 m - kgs (8 to 9.5 ft lbs) |
| Crankshaft pinion nut | 4.0 to 4.5 m - kgs (29 - 32.5 ft lbs) |
| Front wheel spindle nut | 4.0 to 4.5 m - kgs (29 to 32.5 ft lbs) |
| Front fork pinch bolts | 1.6 to 2.4 m - kgs (11.6 to 17 ft lbs) |
| Engine mounting bolts | 1.4 to 2.2 m - kgs (10 to 16 ft lbs) |
| Swinging arm pivot shaft | 3.0 to 4.8 m - kgs (21.7 to 35 ft lbs) |
| Rear wheel spindle nut | 4.0 to 5.0 m - kgs (29 to 36 ft lbs) |
| Rear wheel sprocket bolts | 1.4 to 2.2 m - kgs (10 to 17 ft lbs) |

# Recommended lubricants

| Component | Grade | Quantity |
|---|---|---|
| Engine | SAE 30 two-stroke oil | As required |
| Gearbox | SAE 10W/30 motor oil | 700 cc (1.48/1.23 US/Imp pints) |
| Front forks | Fork oil | 150 cc (0.31/0.26 US/Imp pints) |
| Final drive chain | Chain lubricant | As required |
| Greasing points | High melting point grease | As required |

# Chapter 1 Engine, clutch and gearbox

*For information relating to 1977 on models refer to Chapter 7*

### Contents

| | |
|---|---|
| General description ... 1 | Engine casings and covers: examination and renovation ... 23 |
| Operations with engine in frame ... 2 | Engine reassembly: general ... 24 |
| Operations with engine removed ... 3 | Engine reassembly: gear clusters and selector mechanism: reassembly and replacement ... 25 |
| Method of engine/gearbox removal ... 4 | Engine reassembly: replacing the gearbox components ... 26 |
| Removing engine/gearbox unit ... 5 | crankshaft shim ... 26 |
| Dismantling the engine and gearbox unit: general ... 6 | Preparation of crankcase jointing surfaces: joining the crankcases ... 27 |
| Dismantling the engine and gearbox unit: removing the cylinder head, barrel and piston ... 7 | Reassembly of the gearchange, kickstart mechanism and clutch ... 28 |
| Dismantling the engine: removing the generator ... 8 | Replacing the gearbox sprocket and generator ... 29 |
| Dismantling the engine: removing the left-hand engine casing fittings ... 9 | Engine reassembly: replacing the piston and rings ... 30 |
| Dismantling the engine: removing the right-hand engine casing fittings ... 10 | Engine reassembly: replacing the cylinder barrel ... 31 |
| Dismantling the engine: removing the clutch unit ... 11 | Engine reassembly: replacing the cylinder head ... 32 |
| Dismantling the engine: removing the gearchange and kickstart mechanisms ... 12 | Refitting the engine/gear unit in the frame ... 33. |
| Dismantling the engine: separating the crankcase halves ... 13 | Engine reassembly: reconnecting the carburettor, fuel, oil, and drain hoses ... 34 |
| Dismantling the engine: removing the crankshaft assembly and gear clusters ... 14 | Engine reassembly: reconnecting, bleeding and setting the oil pump ... 35 |
| Examination and renovation: general ... 15 | Engine reassembly: replacing the left-hand outer casing ... 36 |
| Crankshaft assembly: examination and replacement ... 16 | Engine reassembly: replacing the exhaust pipe and silencer ... 37 |
| Small end bearings: examination and renovation ... 17 | Engine reassembly: final adjustments ... 38 |
| Piston and piston rings: examination and renovation ... 18 | Starting and running the rebuilt engine ... 39 |
| Cylinder barrel: examination and renovation ... 19 | Fault diagnosis: engine ... 40 |
| Cylinder head: examination and renovation ... 20 | Fault diagnosis: clutch ... 41 |
| Gearbox components: examination and renovation ... 21 | Fault diagnosis: gearbox ... 42 |
| Clutch assembly: examination and renovation ... 22 | |

### Specifications

**Engine (general)** | **RS100** | **RS125**
---|---|---
Type | Single cylinder air-cooled two-stroke |
Porting | Seven port, torque induction |
Capacity | 97 cc | 123 cc
Bore | 52 mm | 56 mm
Stroke | 45.6 mm | 50 mm
Compression ratio | 6.7 : 1 | 7 : 1
Lubrication | By separate oil pump - Yamaha Autolube system |

**Piston**
Type | Light alloy, two rings |
Clearance in bore | 0.025 — 0.030 mm (0.0010 — 0.0012 in) | 0.030 — 0.035 mm (0.0012 — 0.0014 in)

**Piston rings**
Type | Keystone, pegged |
End gap | 0.15 - 0.35 mm (0.006 - 0.014 in) |

## Chapter 1: Engine, clutch and gearbox

| | RS100 | RS125 |
|---|---|---|
| **Clutch** | | |
| Type | Wet, multi plate | Wet, multi plate |
| No. of plain plates | 5 | 5 |
| No. of friction plates | 4 | 4 |
| No. of springs | 5 | 5 |
| Spring free length | 34.0 mm (1.34 in) | 31.5 mm (1.24 in) |
| Wear limit | 33.0 mm (1.30 in) | 30.5 mm (1.20 in) |
| Friction plate thickness | 4.0 mm (0.157 in) | 3.2 mm (0.126 in) |
| Wear limit | 3.7 mm (0.146 in) | 2.9 mm (0.114 in) |
| **Gearbox** | | |
| Type | Constant mesh, 5-speed | |
| **Gearbox ratios** | | |
| Bottom gear | 2.833 : 1 | 2.833 : 1 |
| 2nd gear | 1.875 : 1 | 1.875 : 1 |
| 3rd gear | 1.368 : 1 | 1.368 : 1 |
| 4th gear | 1.091 : 1 | 1.091 : 1 |
| Top gear | 0.957 : 1 | 0.957 : 1 |
| Secondary reduction ratio | 2.250 : 1 | 2.333 : 1 |
| Primary drive | gear | gear |
| Secondary drive | chain | chain |
| Starting system | Primary kickstart | Primary kickstart |

## 1 General description

The engine unit fitted to the RS100/125 models is an inclined single cylinder two-stroke, employing reed-valve induction for improved tractability. The use of the reed-valve, or torque-induction system, allows more precise induction timing than is possible with conventional two-stroke engines, where the intake and exhaust timing is effected by the covering and uncovering of the relevant ports by the piston. The piston is fitted with two rings of the Keystone type. These rings have a taper across their section which lessens the tendency of the rings to stick or 'gum' in their grooves. The ring ends are located by pegs, as in all two-stroke engines, to obviate the risk of the ring ends catching in the port windows and fracturing.

The engine and gearbox are of unit construction, the crankshaft assembly and gearbox internals being supported by common, vertically split, castings.

Two-stroke engines rely to some degree on secondary, or crankcase compression, and consequently the crankcase must be effectively sealed from both the outside air and from the gearbox. It will be noticed that whilst the gearbox is built in unit with the engine, it is effectively a separate assembly contained inside its own area within the engine casings.

The crankshaft assembly is effectively isolated by the use of large diameter oil seals.

The flywheel generator is mounted on the left-hand side of the engine, just outboard of the crankcase. The generator coils are mounted on a stator plate beneath the flywheel.

The clutch is mounted on the right-hand side of the engine unit, drive being taken to the outside of the clutch drum via a helical gear. An outer cover is fitted over the clutch and primary gears, providing an oil bath for both.

The carburettor is mounted on a rubber inlet stub at the rear of the cylinder. The exhaust gases are expelled through an exhaust pipe and silencer.

The gearchange pedal is mounted on the left-hand side of the machine, selecting the five constant-mesh ratios.

Lubrication is effected by the Yamaha Autolube system, which takes the form of a gear driven oil pump located in the front of the right-hand outer casing. Oil drawn from a separate oil tank is fed directly to the various engine internals. The pump is controlled by the throttle twistgrip, thus metering exactly the lubrication requirements of the engine under varying loads. This system eliminates the need for premixed two-stroke fuel, which is at best something of a compromise.

## 2 Operation with engine in frame

It is not necessary to remove the engine unit from the frame unless the crankshaft assembly and/or the gearbox internals require attention. Most operations can be accomplished with the engine in place, such as:
1. Removal and replacement of the cylinder head.
2. Removal and replacement of the cylinder barrel and piston.
3. Removal and replacement of the generator.
4. Removal and replacement of the clutch.
5. Removal and replacement of the contact breaker assembly.

When several operations need to be undertaken simultaneously, it will probably be advantageous to remove the complete engine unit from the frame, an operation that should take approximately two hours, working at a leisurely pace. This will give the advantage of better access and more working space.

## 3 Operations with engine removed

1. Removal and replacement of the crankshaft assembly.
2. Removal and replacement of the gear cluster, selectors and gearbox main bearings.

## 4 Method of engine/gearbox removal

As mentioned previously, the engine and gearbox are of unit construction, and it is necessary to remove the unit complete, in order to gain access to the internal components. Separation and reassembly are only possible with the engine unit removed from the frame. It is recommended that the procedure detailed below is adhered to, as in certain instances components are much easier to remove whilst the unit is supported by the frame. This applies particularly to the gearbox drive sprocket nut, which is best slackened before the final drive chain is removed.

## Chapter 1: Engine, clutch and gearbox

### 5 Removing the engine/gearbox unit

1 Place the machine on the centre stand and make sure that it is standing firmly on level ground.

2 Turn the fuel tap to the 'stop' position. Disconnect the fuel pipe where it joins the carburettor float chamber. The pipe is pushed over a stub and retained by a small wire clip. The clip can be moved after squeezing the ends together and the pipe can then be eased off the stub, using a small screwdriver if necessary.

The tank can now be removed and placed in a safe place. Any remaining fuel must be drained off first, and the pipe between the tank halves disconnected. Lift the rear of the tank and pull gently backwards, having first lifted the dualseat. The rear of the tunnel in the underside of the tank rests on a rubber stop, the front being retained by two rubber buffers to the rear of the headstock.

With the tank removed, better access is gained, and the tank's paintwork is less likely to sustain damage. Remove the oil tank, having first drained off the contents, or plugged the feed pipe to contain the oil. The tank is retained by two screws. Prise off the connecting rubber tube between the carburettor and air filter trunking.

3 Remove the three crosshead screws retaining the generator cover and lift the cover away. If at all possible, an impact driver should be used on these, and all the other casing screws. They may be very hard to move by hand.

4 Slacken the pinch bolt on the gearchange pedal, and lift it clear of its splines.

5 Remove the screws holding the left-hand cover, and lift it away. Note that the clutch cable can now be freed by sliding its trunnion out of the operating arm, via the slot provided, to clear the cable.

6 Slacken the nut holding the exhaust pipe to the silencer, using a C spanner, if possible. The nut can be loosened in the absence of the correct spanner by judicious use of a cold chisel or punch and a hammer, bearing in mind the risk of damaging the nut. Remove the two nuts holding the exhaust pipe flanges to the cylinder barrel and pull the exhaust pipe clear of the machine. The silencer can be left in position at this stage.

7 Pull out the wiring connector situated between the battery and the intake trunking and separate the cable.

8 Detach the tachometer drive cable by unscrewing the gland nut at the front of the right-hand engine casing. Pull the cable clear and secure it so that it will not interfere with engine removal. (Note: tachometer fitted to DX model only).

9 Unscrew the carburettor top and pull upwards, freeing the slide and needle. These need not be removed from the cable, but should be positioned out of harm's way. Similarly, the high tension lead to the spark plug should be detached and tied clear.

10 Remove the front section of the right-hand outer casing after unscrewing the three crosshead screws which retain it. The oil pump will now be exposed, and the operating cable should be detached and withdrawn. This operation is made easier by turning the pulley in which the cable runs so that the cable becomes slack.

11 Remove the various oil feed, drain, and breather pipes from the top of the engine casings and from the carburettor. The order and positions should be noted, to facilitate reassembly. The neutral indicator light lead must also be removed from the left-hand casing.

12 Slacken and remove the gearbox sprocket retaining nut after knocking back the locking tab with a cold chisel. The sprocket may be prevented from turning during this operation by applying the rear brake.

13 The final drive chain can now be removed by displacing the spring clip with a pair of long nosed pliers. Remove the joining link plate and disengage the chain from one of the two pins. Reassemble the joining link on one end of the chain to avoid subsequent loss. Remove the gearbox sprocket.

14 Remove the kickstart pedal after slackening the pinch bolt. Slacken and remove the nut on the end of the swinging arm spindle and disengage the silencer. Remove the nut retaining the silencer to its lower mounting bracket and remove and place the silencer where it will not get damaged.

15 The engine/gearbox unit is now retained only by the mounting bolts. Note that they are of differing lengths, and when removing them, lay them out to facilitate reassembly. Remove the nut on the lower of the two rear mounting bolts. This also retains the footrest and prop stand assembly, which must be removed to clear the engine unit. The assembly is also located to the frame by a single short bolt, adjacent to the lower rear engine bolt, which should be removed. It can now be pulled clear of the frame.

Remove the front mounting bolt and the nuts from the two rear mounting bolts, leaving the engine suspended by the rear bolts only.

16 The engine/gearbox unit can be lifted easily by one person only, but it is better if a second person is at hand to withdraw the final two bolts and steady the machine as the unit is pulled clear. It is slightly easier to remove the unit from the left-hand side of the machine. Lift the unit gently away, ensuring that no pipes or cables have been inadvertently left connected. Place the unit on a suitable workbench to await further dismantling.

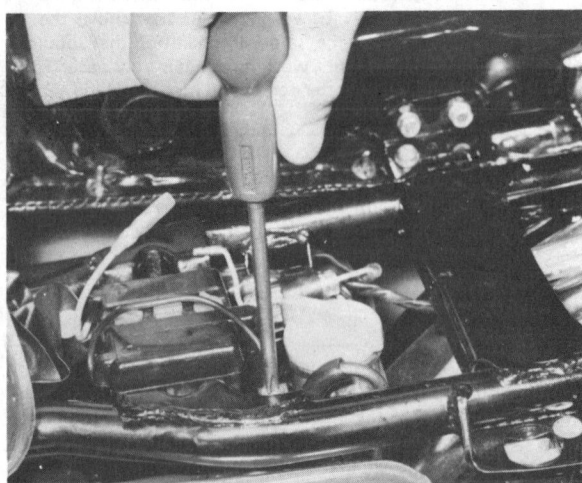

5.2a Remove oil tank after removing mounting screws

5.2b Disconnect intake trunking from carburettor and filter

5.5 Remove inner left-hand cover and detach clutch cable

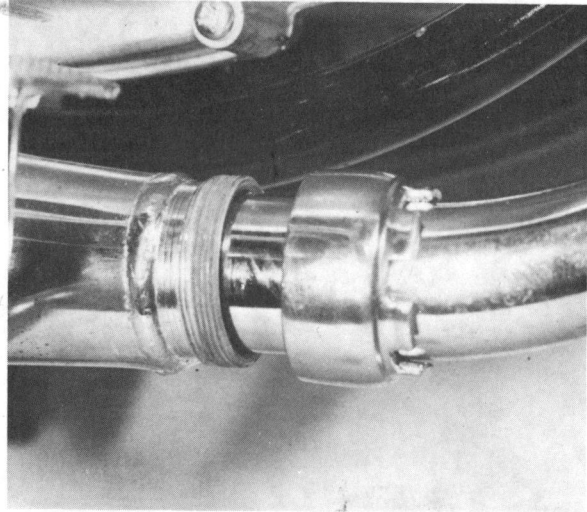
5.6a Slacken exhaust pipe gland nut ...

5.6b ... and flange mounting nuts

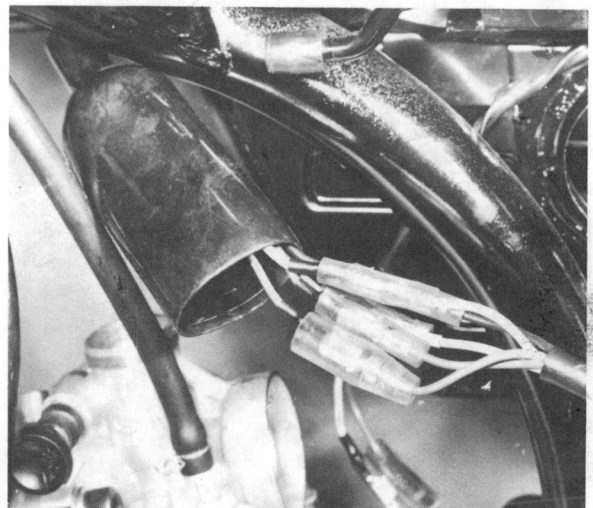
5.7 Separate wiring connectors

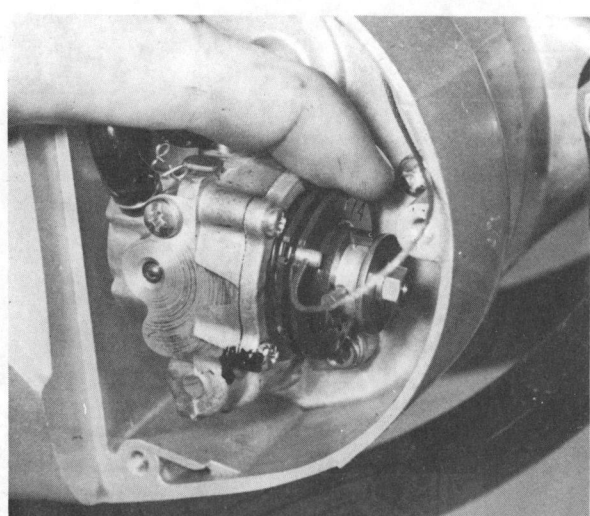
5.10b Detach oil pump control cable from pulley

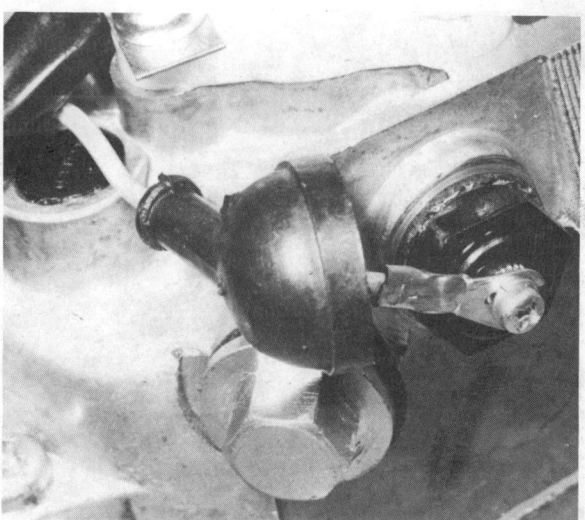
5.11 Detach neutral switch lead

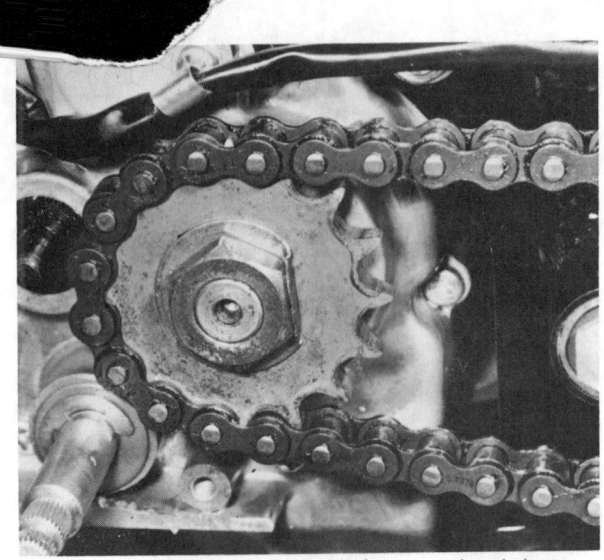
5.12 Slacken gearbox sprocket nut before removing chain

5.14a Kickstart pedal is retained by pinch bolt

5.14b Note front silencer mounting nut

5.15a Remove bolt and lower engine mounting bolt ...

5.15b ... to free footrest assembly from frame

5.16 Engine unit can be lifted from frame

## Chapter 1: Engine, clutch and gearbox

### 6 Dismantling the engine and gearbox unit: general

1  Before commencing work on the engine unit, the external surfaces should be cleaned thoroughly. A motor cycle engine has very little protection from road grit and other foreign matter, which will find its way into the dismantled engine if this simple precaution is not taken. One of the proprietary cleaning compounds, such as 'Gunk' or 'Jizer' can be used to good effect, particularly if the compound is permitted to work into the film of oil and grease before it is washed away. Special care is necessary when washing down to prevent water from entering the now exposed parts of the engine unit.
2  Never use undue force to remove any stubborn part unless specific mention is made of this requirement. There is invariably good reason why a part is difficult to remove, often because the dismantling operation has been tackled in the wrong sequence.
3  Mention has already been made of the benefits of owning an impact screwdriver. Most of these tools are equipped with a standard ½ inch drive and an adaptor which can take a variety of screwdriver bits. It will be found that most engine casing screws will need jarring free due to both the effects of assembly by power tools and an inherent tendency for screws to become pinched in alloy castings.

A cursory glance over many machines of only a few years use, will almost invariably reveal an array of well-chewed screw heads. Not only is this unsightly; it can also make emergency repairs impossible. It should also be borne in mind that there are a number of types of crosshead screwdrivers which differ in the angle and design of the driving tangs. To this end, it is always advisable to ensure that the correct tool is available to suit a particular screw.
4  In addition to the above points, it is worth noting before any dismantling work is undertaken, that some of the screws holding the inner engine casings together are recessed deeply between cast webs. These screws may require the use of an impact screwdriver fitted with an extended bit.

### 7 Dismantling the engine and gearbox unit: removing the cylinder head, barrel and piston

1  The carburettor is retained by two socket screws to the reed-valve case. It can be removed after the screws have been detached, or alternatively, the carburettor and valve case can be removed as a unit by unscrewing the four reed-valve case retaining bolts.
2  The cylinder head is retained by four sleeve bolts. Slacken them in a diagonal sequence and then carefully remove the cylinder head and gasket.
3  Lift the cylinder barrel carefully up along the studs, taking care to catch the piston as it emerges from the bore. As a precaution against pieces of broken piston ring entering the crankcase mouth, a piece of clean rag should be used to pack around the connecting rod, covering the opening.
4  Remove the circlips from the piston and press out the gudgeon pin so that the piston is released. If the gudgeon pin is a particularly tight fit, the piston should be warmed first, to expand the alloy and release the grip on the steel pin. If it is necessary to tap the gudgeon pin out of position, make sure that the connecting rod is supported to prevent distortion. On no account use excess force. Discard the old circlips.
5  The small end bearing consists of needle rollers running in a cage. With the gudgeon pin removed, it can be displaced easily from the connecting rod eye.

### 8 Dismantling the engine: removing the generator

1  The generator flywheel is keyed to the crankshaft and is retained by a central securing nut. It is necessary to prevent crankshaft rotation in order to slacken the flywheel nut. If this is being done with the unit in the frame, the machine should be placed in top gear and the rear brake applied. If, however, the generator is being removed in the course of dismantling the engine unit, the crankshaft can be locked by passing a bar through the connecting rod eye and resting the ends on wooden blocks placed at either side of the crankcase mouth.
2  With the crankshaft effectively locked, the flywheel nut can be removed and the flywheel drawn off. If the official Yamaha extractor is used, it should be noted that this has a **left-hand thread**. Alternatively, a two-legged puller may be used in place of the official extractor, providing a discarded nut or similar is used as a spacer between the centre bolt of the puller and the crankshaft end to prevent damage to the latter. Tighten the centre bolt gently and tap the end to dislodge the flywheel rather than use excessive force. Remove the flywheel and place it to one side, together with its Woodruff key.
3  Remove the two countersunk screws which retain the generator stator to the crankcase and draw the stator assembly away, disengaging the output leads from the inner casing guide. Note that there is no need to mark the position of the stator in relation to the crankcase, as it is not adjustable and therefore cannot be reassembled incorrectly.

7.2 Cylinder head is retained by four sleeve nuts

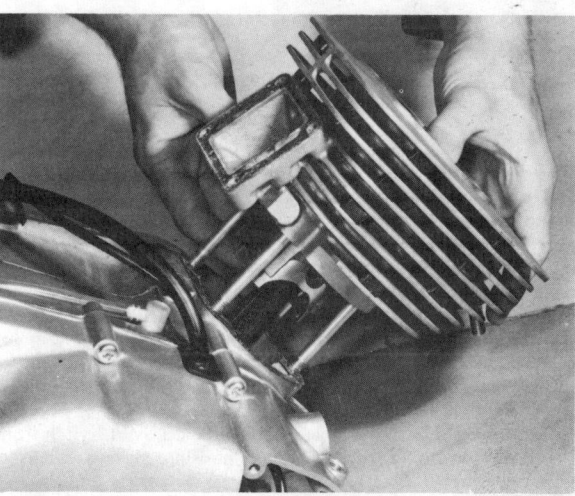

7.3 Support piston as cylinder is lifted off studs

## Chapter 1: Engine, clutch and gearbox

7.4 Pad crankcase mouth before removing circlips

8.2a Slacken flywheel retaining nut ...

8.2b ... and draw off the generator flywheel

8.3 Stator is retained by two countersunk screws

### 9 Dismantling the engine: removing the left-hand engine casing fittings

1  Remove the three screws securing the gear selector drum end cap and then the cap. The cap holds in position a horseshoe-shaped retainer, fitted in a groove in the end of the selector drum. This will drop away when the cap is removed.
2  Withdraw the clutch pushrod, checking to see that the ball between the two pushrods is not lost. It may remain inside the casing and if so should be shaken out.
3  Unscrew and withdraw the neutral indicator light switch from the top of the casing, also the plunger housing, spring and ball assembly situated directly behind it.
4  Remove the circlip from the gearchange pedal shaft and check finally that nothing has been missed before turning the unit over to work on the right-hand casing.

9.1a Remove three cross-headed screws ...

## Chapter 1: Engine, clutch and gearbox

9.1b ... and lift off plastic cover to free retainer

9.3a Remove neutral indicator switch ...

9.3b ... followed by detent ball and spring assembly

9.4 Crossover shaft is retained by clip and washer

### 10 Dismantling the engine: removing the right-hand engine casing and fittings

1  With the unit supported on suitable blocks, slacken and remove the outer cover screws, and lift the cover away, complete with the oil pump and drive gear.

2  Lift off the tachometer plastic drive gear and washers from its support pin (where fitted).

### 11 Dismantling the engine: removing the clutch unit

1  Unscrew the five clutch tension bolts diagonally and remove, complete with springs. If the clutch cannot easily be held by hand, a small wad of rag may be wedged between the drum and driving gear to lock it in position.

2  Lift out the clutch driven and drive plates, noting the anti-drag rubber 'O' rings fitted between the plates. Withdraw the mushroom headed pushrod.

3  To remove the clutch centre nut, the centre will have to be locked. This can be accomplished by placing the gearbox sprocket temporarily on its splines and holding it, either with the final drive chain bunched up so that it lodges against the casting, or by using a suitable chain wrench.

4  If it is envisaged that this operation will be repeated a number of times during ownership, the correct holding tool is easily fabricated by brazing or welding about one foot of ¼ inch steel rod to the edge of a discarded plain clutch plate. This simple tool can then be used to hold the clutch centre during the centre nut slackening and tightening procedure. A similar device is available as a Yamaha service tool.

5  If using the final drive chain or a chain wrench on the gearbox sprocket, top gear should be selected as this will place less strain on the gear teeth. Be especially careful not to exert excessive pressure. This should not be necessary, and there is risk of damage to the alloy casing where the drive chain method has been utilised.

6  Lift away the clutch centre and thrust washer. Remove the gearbox sprocket, if used as described above.

7  Before removing the clutch outer drum, the crankshaft pinion nut should be slackened off. This is done after the pinion and crank assembly has been locked to prevent it turning when pressure is applied to the nut. A small wad of rag can be placed between the pinion and clutch drum. This forms an effective wedge without unduly overstressing the teeth. A more satisfactory method is to lock the crank assembly by placing a short metal rod through the small end eye of the connecting rod, and supporting the two ends of the rod on wooden blocks placed each side of the crankcase mouth.

# Chapter 1: Engine, clutch and gearbox

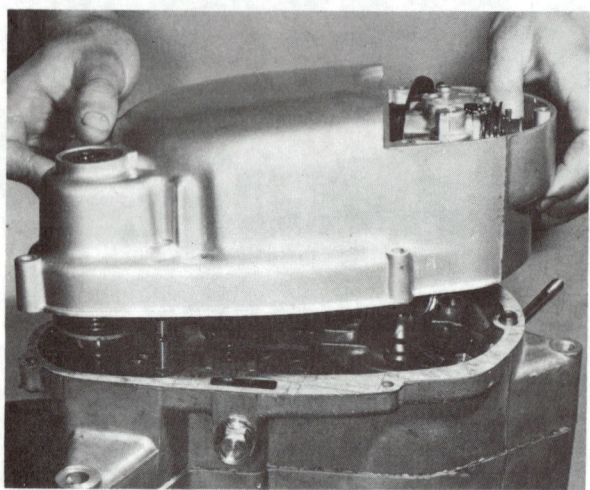

10.1 Remove right-hand outer cover with pump

10.2 Lift off tachometer drive pinion where fitted

11.7 Crankshaft may be locked as shown

## 12 Dismantling the engine: removing the gearchange and kick-start mechanisms

1  Disengage the forked end of the gearchange linkage from the selector pins at the end of the selector drum. Pull the assembly away, complete with gearchange shaft which runs through the casings. Unscrew the single bolt which retains the ratchet pawl and lift it away, disengaging the spring from the retainer plate.
2  Unhook the kickstart return spring from its locating peg and release the tension. This can safely be done by hand - the assembly is not under great pressure. The release spring will unwind only by about 180°.
3  Withdraw the kickstart mechanism as an assembly. Remove the circlip from the idler gear, using circlip pliers, and slide off the gear and spring. Slide the clutch bearing sleeve from its shaft.
4  Remove the circlip retaining the kickstart idle pinion and slide off the pinion and washer. Remove the crankshaft pinion, taking care not to lose the Woodruff key.

## 13 Dismantling the engine: separating the crankcase halves

1  The engine/gearbox casing halves are retained by a total of eleven crosshead screws passing through from the left-hand casing.
2  Slacken and remove the casing screws, noting that some are deeply recessed between the cast webs, and may require the use of an impact driver with an extended bit. Check that nothing can impair separation of the casings.
3  Support the unit on blocks, right-hand side uppermost, and draw the two halves apart. To aid separation, the casings may be jarred apart using a hammer interspersed with a wooden block. On no account use a hammer directly on the alloy casings as the material is brittle and is easily fractured.
4  If the cases still refuse to part, do not use excessive force, but check that separation is not impeded by a forgotten screw or component.
5  As the two halves part, make sure that the gear cluster and selector drum remain in the left-hand casing.

## 14 Dismantling the engine: removing the crankshaft assembly and gear clusters

1  With the casing halves supported on a bench, the gear clusters may be lifted away, complete with the selector drum and forks. Place the assembly to one side, to await further attention. If no further attention is felt necessary, it is worthwhile securing the assembly with two rubber bands. This will prevent any component becoming displaced.
2  The crankshaft assembly is now retained by the bearing of one casing. It is invariably a tight fit, and whilst it is possible to drive it out using a hide mallet, care should be exercised. The manufacturers advocate the use of two service tools to facilitate this operation, one being the crankcase separation tool, and the other a sleeve puller for reassembly.
3  Crankshaft reconditioning, particularly replacing the bearing, will necessitate crankshaft separation and realignment to a high degree of accuracy. If, therefore, the crankshaft assembly is in need of attention, it is best left to a Yamaha dealer, who will have the various tools and equipment to hand. If this is the case, then it is probably safer to hand the crankshaft assembly, in position in the crankcase, to the dealer for his attention.
4  If, for any other reason, it is decided that the crankshaft must be removed, the case must first be securely supported on wooden blocks to distribute the load evenly. Place a hardwood block on the end of the mainshaft to avoid burring the end, and jar the assembly free, using a hide mallet.

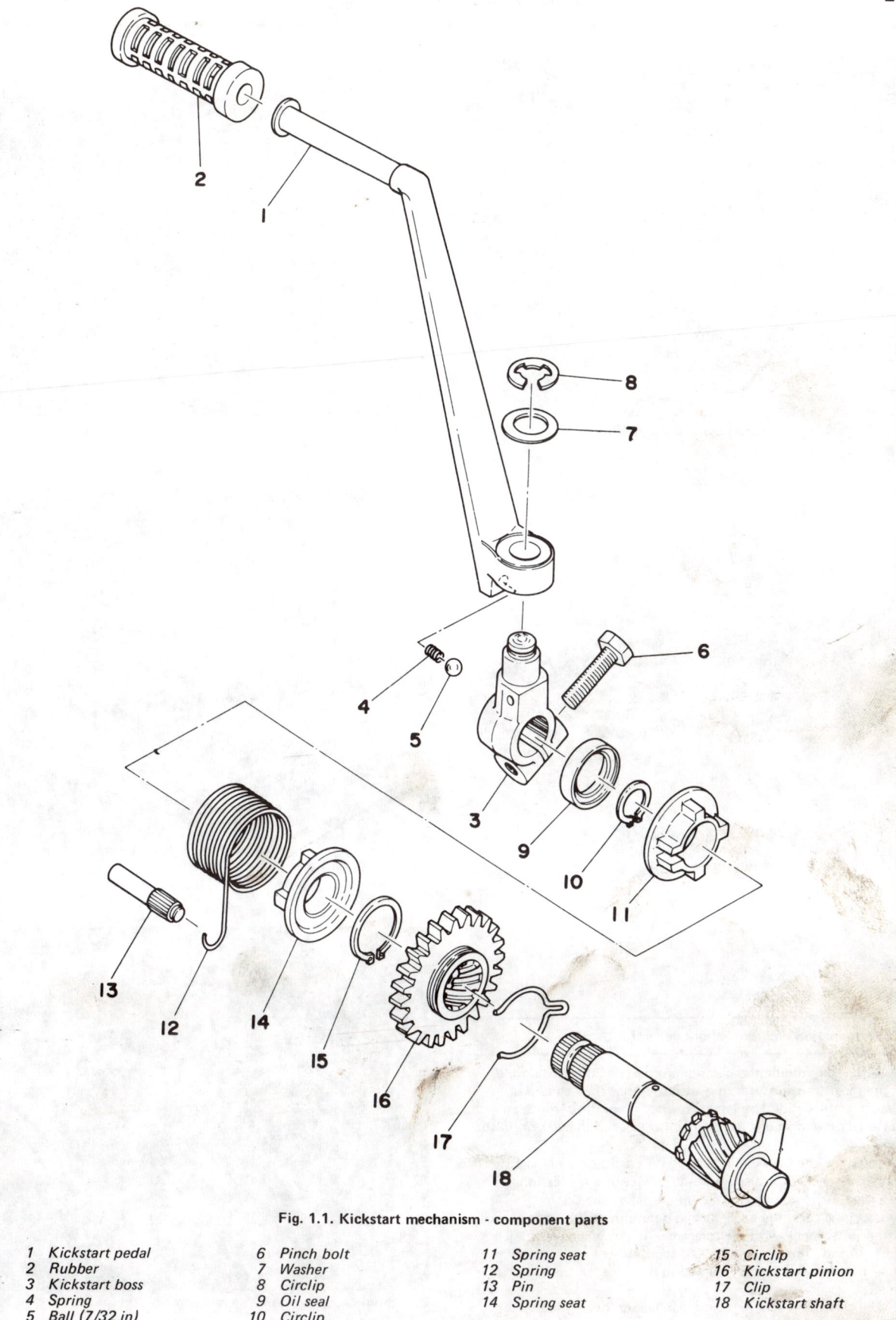

Fig. 1.1. Kickstart mechanism - component parts

1 Kickstart pedal
2 Rubber
3 Kickstart boss
4 Spring
5 Ball (7/32 in)
6 Pinch bolt
7 Washer
8 Circlip
9 Oil seal
10 Circlip
11 Spring seat
12 Spring
13 Pin
14 Spring seat
15 Circlip
16 Kickstart pinion
17 Clip
18 Kickstart shaft

Chapter 1: Engine, clutch and gearbox

12.1 Remove countershaft assembly and stopper

13.5 Make sure crankcase components remain in LH case

14.1 Lift out gearbox internals

14.4 Crankshaft assembly can be driven out of casing

## 15 Examination and renovation: general

1  Before examining the component parts of the dismantled engine/gear unit for wear, it is essential that they should be cleaned thoroughly. Use a paraffin/petrol mix to remove all traces of oil and sludge which may have accumulated within the engine.
2  Examine the crankcase castings for cracks or other signs of damage. If a crack is discovered, it will require professional attention, or in an extreme case, renewal of the casting.
3  Examine carefully each part to determine the extent of wear; if in doubt, check with the tolerance figures whenever they are quoted in the text. The following sections will indicate what type of wear can be expected and in many cases, the acceptable limits.
4  Use clean, lint-free rags for cleaning and drying the various components, otherwise there is a risk of small particles obstructing the internal oilways.

15.2 Crankcase should be examined for cracks

## Chapter 1: Engine, clutch and gearbox

### 16 Crankshaft assembly - examination and replacement

1 The crankshaft assembly comprises a pair of flywheels between which is the connecting rod supported by its crankpin and bearing. Mainshafts run from each flywheel centre and are carried in main bearings in the crankcase halves. The assembly forms a pressed-up unit.

2 In the event of big-end failure, it is beyond the means of the average owner to separate the flywheel assembly and to realign it to the high standard of accuracy required. In consequence the complete crankshaft assembly must be taken to a Yamaha specialist for the necessary repairs and renovation, or service exchanged for a fully reconditioned unit.

3 Main bearing failure will immediately be obvious when the bearings are inspected, after the old oil has been washed out. If any play is evident or if the bearings do not run freely, renewal is essential. Warning of main bearing failure is usually given by a characteristic rumble that can be readily heard when the engine is running. Some vibration will also be felt, which is transmitted via the footrests.

4 Big-end failure is characterised by a pronounced knock that will be most noticeable when the engine is working hard. There should be no play whatsoever in the connecting rod, when it is pushed and pulled in a vertical direction. A small amount of sideways play is permissible, but not more than 0.3 mm (0.012 in).

5 Oil seal failure is a common occurrence in two-stroke engines that have seen a reasonable amount of service. When the oil seals begin to wear, air is admitted to the crankcase which will dilute the incoming mixture. This in turn causes uneven running and difficulty in starting.

6 The oil seals at each end of the crankshaft are easy to renew when the engine is stripped; they are a push fit over each end of the crankshaft adjacent to the outer main bearings. It is a wise precaution to renew these seals whenever the engine is stripped, irrespective of their condition.

### 17 Small end bearing: examination and renovation

1 The small end bearing is a caged needle roller and will seldom give trouble unless a lubrication failure has occurred. The gudgeon pin should be a good sliding fit in the bearing, without any play. If play develops a noticeable rattle will be heard when the engine is running, indicative of the need for renewal of the bearings.

2 No problem is encountered when replacing the caged needle roller bearings as they are a light push fit in the eye of the connecting rod. New small end bearings are normally supplied whenever the crankshaft assembly is renewed or service-exchanged.

### 18 Piston and piston rings: examination and renovation

1 Attention to the piston and rings can be overlooked if a rebore is necessary because a new piston and rings will be fitted under these circumstances.

2 If a rebore is not considered necessary, the piston should be examined closely. Reject the piston if it is badly scored or discoloured as the result of the exhaust gases by-passing the rings. Check the gudgeon pin bosses to ensure they are not enlarged or that the grooves retaining each circlip are not damaged.

3 Remove all carbon from the piston crown and use metal polish to finish off, so that a high polish is obtained. Carbon will adhere much less readily to a polished surface. Examination of the piston crown will show whether the engine has been rebored previously since the amount of overbore is invariably stamped on each piston crown. Two oversizes are available, namely 0.25 and 0.50 mm.

4 Remove the piston rings by pushing the ends apart with the thumbs whilst gently easing each ring from its groove. Great care is necessary throughout this operation because the rings are brittle and will break easily if overstressed. If the rings are gummed in their grooves, three strips of tin can be used, to ease them free, as shown in the accompanying illustration.

5 Piston ring wear can be checked by inserting the rings one at a time in the cylinder bore from the top and pushing them down about 1½ inches with the base of the piston so that they rest squarely in the bore. Make sure that the end gap is away from any of the ports. If the end gap is within the range 0.15 - 0.35 mm (0.006 - 0.014 in) the ring is fit for further service.

6 Examine the working surface of each piston ring. If discoloured areas are evident, the ring should be renewed because these areas indicate the blow-by of gas. Check that there is not a build-up of carbon on the back of the ring or in the piston ring groove, which may cause an increase in the radial pressure. A portion of broken ring affords the best means of cleaning out the piston ring grooves.

7 Check that the piston ring pegs are firmly embedded in each piston ring groove. It is imperative that these retainers should not work loose, otherwise the rings will be free to rotate and there is danger of the ends being trapped in the ports.

8 It cannot be over-emphasised that the condition of the piston and piston rings is of prime importance because they control the opening and closing of the ports by providing an effective moving seal. A two-stroke engine has only three working parts, of which the piston is one. It follows that the efficiency of the engine is very dependent on the condition of the piston and the parts with which it is closely associated.

17.1 Check condition of gudgeon pin and bearing

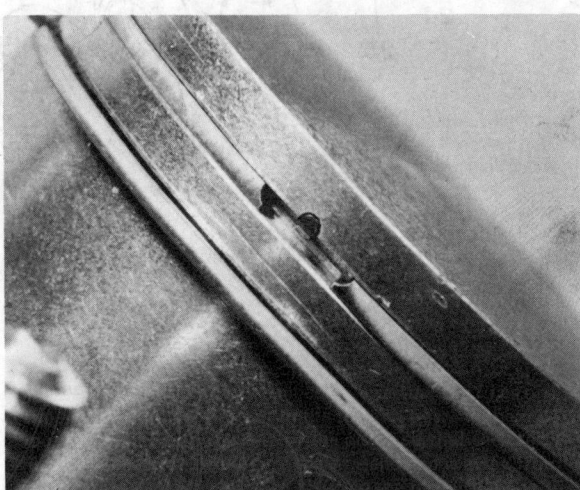

18.7 Note retaining pins in piston ring grooves

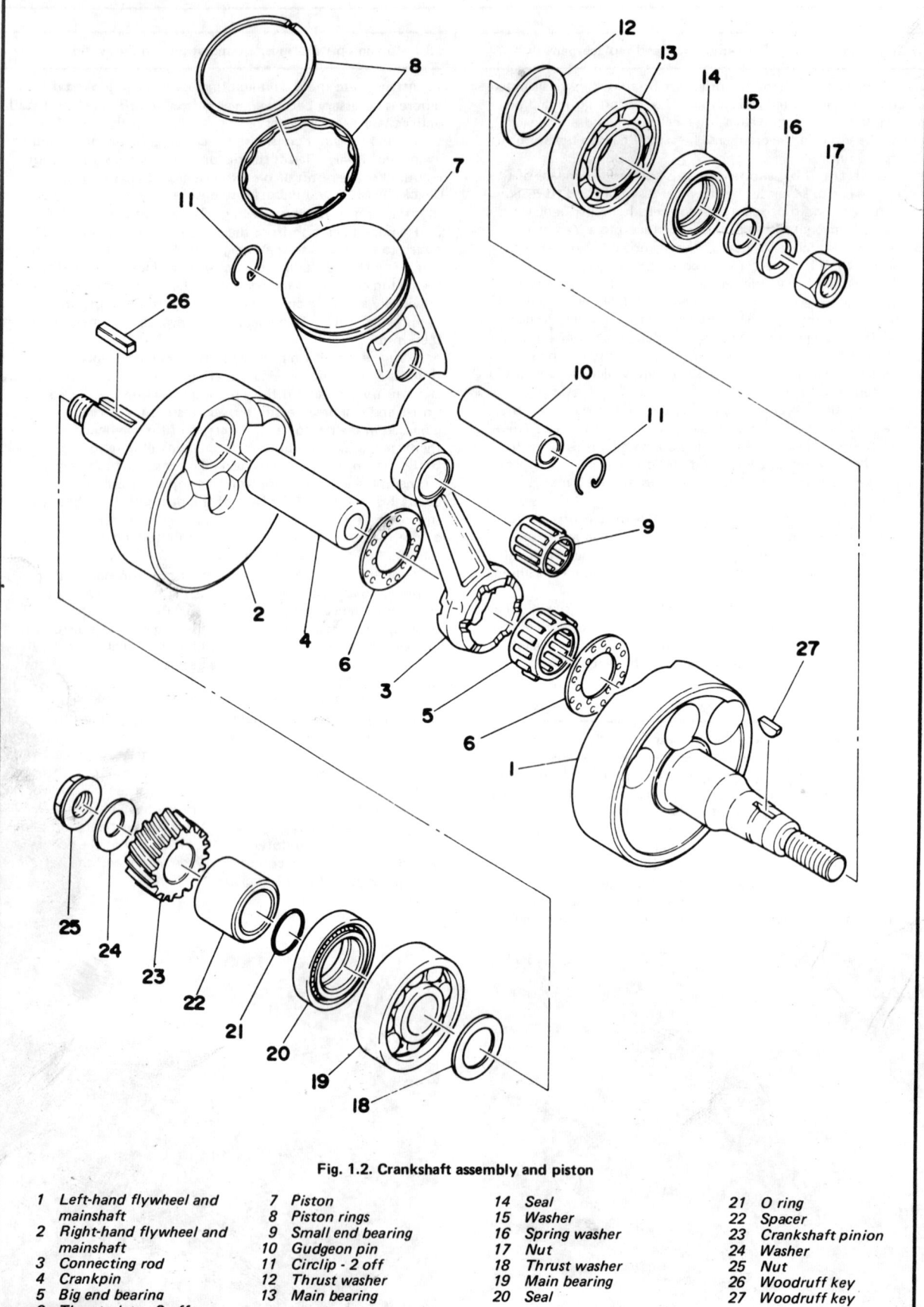

Fig. 1.2. Crankshaft assembly and piston

1 Left-hand flywheel and mainshaft
2 Right-hand flywheel and mainshaft
3 Connecting rod
4 Crankpin
5 Big end bearing
6 Thrust plate - 2 off
7 Piston
8 Piston rings
9 Small end bearing
10 Gudgeon pin
11 Circlip - 2 off
12 Thrust washer
13 Main bearing
14 Seal
15 Washer
16 Spring washer
17 Nut
18 Thrust washer
19 Main bearing
20 Seal
21 O ring
22 Spacer
23 Crankshaft pinion
24 Washer
25 Nut
26 Woodruff key
27 Woodruff key

# Chapter 1: Engine, clutch and gearbox

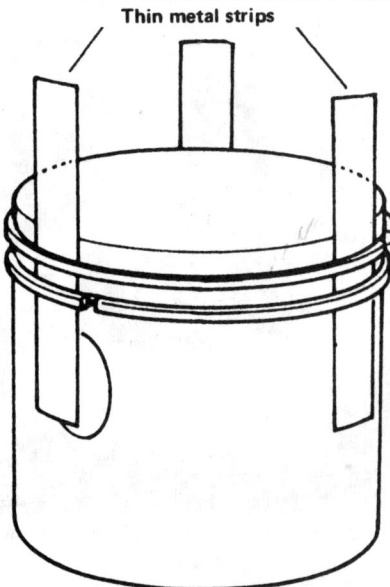

**Fig. 1.3. Freeing gummed rings**

### 19 Cylinder barrel: examination and renovation

1  There will probably be a lip at the uppermost end of the cylinder bore which marks the limit of travel of the top of the upper piston ring. The depth of the lip will give some indication of the amount of bore wear that has taken place even though the amount of wear is not evenly distributed.

2  Insert the piston (without rings) in its cylinder bore so that it is about ¾ inch from the top of the bore. Measure the clearance between the piston skirt and the cylinder wall with a feeler gauge. Repeat at two further positions lower down the bore. The recommended clearance is from 0.030 - 0.035 mm (0.0012 - 0.0014 in); if the clearance exceeds 0.1 mm (0.004 in) the cylinder is in need of a rebore.

3  Give the cylinder barrel a close visual inspection. If the surface of the bore is scored or grooved, indicative of an earlier seizure or a displaced circlip and/or gudgeon pin, a rebore is essential. Compression loss has a marked effect on engine performance.

4  Check that the outside of the cylinder barrel is clean and free from road dirt. Use a wire brush on the cooling fins if they are obstructed in any way but take care not to score or damage the light alloy. If the air flow to the cooling fins is obstructed, the engine may overheat badly. Although caustic soda is often recommended for cleaning some of the oilier components of a two-stroke engine, NEVER use it on parts made of light alloy. Caustic soda attacks aluminium alloy with great vigour and produces an explosive gas.

5  Clean all carbon deposits from the exhaust ports using a blunt ended scraper. It is important that all the ports should have a clean, smooth appearance because this will have the dual benefit of improving gas flow and making it less easy for carbon to adhere in the future. Finish off with metal polish, to heighten the polishing effect.

6  Do not under any circumstances enlarge or alter the shape of the ports under the mistaken belief that improved performance will result. The size and position of the ports predetermines the characteristics of the engine and unwarranted tampering can produce very adverse effects.

### 20 Cylinder head: examination and renovation

1  It is unlikely that the cylinder head will require any special attention apart from removing the carbon deposit from the combustion chamber. Finish off with metal polish; the polished surface will help improve gas flow and reduce the tendency of future carbon deposits to adhere so readily.

2  Check that the cooling fins are clean and unobstructed, so that they receive the full air flow. The rubber blocks should not be omitted, as they dampen resonance between the fins, thus reducing engine noise.

3  Check the condition of the thread within the spark plug hole. The thread is easily damaged if the spark plug is overtightened. If necessary, a damaged thread can be reclaimed by fitting a Helicoil thread insert. Most Yamaha agents have facilities for this type of repair, which is not expensive.

4  If there has been evidence of oil seepage from the cylinder head joint when the machine was in use, check whether the cylinder head is distorted by laying it on a sheet of plate glass. Severe distortion will necessitate renewal of the cylinder head, but if the distortion is only slight, the head can be reclaimed by wrapping a sheet of emery cloth around the glass and using it as the surface on which to rub down the head with a rotary motion, until it is once again flat. The usual cause of distortion is failure to tighten down the cylinder head bolts evenly, in a diagonal sequence.

### 21 Gearbox components: examination and renovation

1  Examine each of the gear pinions to ensure that there are no chipped or broken teeth and that the dogs on the end of the pinions are not rounded. Gear pinions with these defects must be renewed; there is no satisfactory method of reclaiming them.

2  Examine the selector forks carefully, ensuring that there is no scoring or wear where they engage in the gears, and that they are not bent. Damage and wear rarely occur in a gearbox which has been properly used and correctly lubricated, unless very high mileages have been covered.

3  The tracks in the selector drum, which co-ordinate the movement of the selector forks, should not show signs of undue wear. Check also that the plunger spring, which bears upon the detent ball, has not weakened, and that no play has developed in the gear selector linkages.

4  Unless the unit has shown signs of wear or malfunctioning, it is unnecessary to dismantle the kickstart assembly. It is, however, a very simple unit to deal with.

5  Check the movement of the gear on the spiral which brings it into engagement with the coupling gear. Look for chipped or broken teeth and excessive wear, replacing the gear if necessary.

6  Remove the circlip retaining the spring cover, and slide the cover, spring and spring guide off the shaft. The gear is retained by a second circlip, together with a shim and spring clip.

7  Examine the components for signs of wear, renewing as necessary. Reassembly is a direct reversal of the dismantling sequence.

21.2 Check condition of gears and selectors ...

28　Chapter 1: Engine, clutch and gearbox

21.3 ... and grooves in selector drum

21.6a Kickstart mechanism can be removed as an assembly

21.6b ... or dismantled as shown

21.6c Spring is not under great tension

21.6d Pinion is retained by circlip to shaft

**22 Clutch assembly: examination and renovation**

1  After a considerable mileage has been covered, the bonded linings of the clutch friction plates will wear down to or beyond the specified wear limit, allowing the clutch to slip.
2  The degree of wear is measured across the faces of the friction material, the nominal, or new, size being 4 mm (0.157 in) on the RS100 model and 3.2 mm (0.126 in) in the case of the RS125. If they show more than 0.3 mm (0.012 in) wear, they should be renewed, even if slipping is not yet apparent.
3  The plain plates should be free from scoring and signs of overheating, which will be apparent in the form of blueing. The plates should also be flat. If more than 0.05 mm (0.002 in) out of true, judder or snatch may result.
4  Measure the uncompressed length of the clutch springs which should be 34 mm on the RS100 model and 31.5 mm on the RS125 (1.34 in and 1.24 in). Should the springs have taken a set of 1 mm (0.04 in) or more they should be renewed.
5  Check the condition of the thrust bearing and washer and the bearing face on which it runs, on the clutch centre. Excessive play or wear will cause noise and erratic operation.

Chapter 1: Engine, clutch and gearbox    29

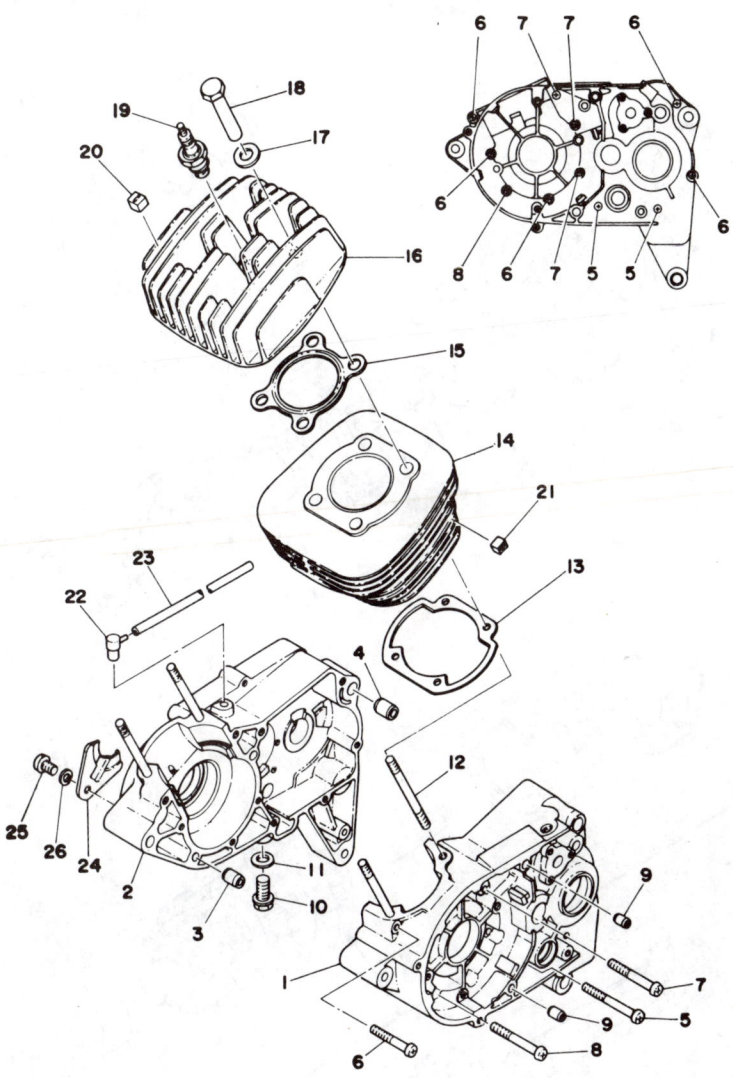

Fig. 1.4. Crankcase, cylinder and cylinder head - component parts

| | | | |
|---|---|---|---|
| 1 Left hand crankcase | 8 Screw | 15 Cylinder head gasket | 21 Damping block - 16 off |
| 2 Right-hand crankcase | 9 Dowel - 2 off | 16 Cylinder head | 22 Breather outlet |
| 3 Dowel | 10 Drain plug | 17 Washer - 4 off | 23 Breather pins |
| 4 Dowel | 11 Fibre washer | 18 Sleeve - 4 off | 24 Retainer |
| 5 Screw - 2 off | 12 Stud - 4 off | 19 Spark plug | 25 Screw |
| 6 Screw - 5 off | 13 Cylinder base gasket | 20 Damping block - 24 off | 26 Spring washer |
| 7 Screw - 3 off | 14 Cylinder | | |

6  Check the condition of the slots in the outer surface of the clutch centre and the inner surfaces of the outer drum. In an extreme case, clutch chatter may have caused the tongues of the inserted plates to make indentations in the slots of the outer drum, or the tongues of the plain plates to indent the slots of the clutch centre. These indentations will trap the clutch plates as they are freed, and impair clutch action. If the damage is only slight the indentations can be removed by careful work with a file and the burrs removed from the tongues of the clutch plates in a similar fashion. More extensive damage will necessitate renewal of the parts concerned.

7  The clutch release mechanism attached to the inside of the left-hand crankcase cover does not normally require attention, provided it is greased at regular intervals. It is held to the cover by two crosshead screws and operates on the worm and quick start thread principle. A light return spring ensures that the pressure is taken from the end of the clutch pushrod when the handlebar lever is released and the clutch fully engaged.

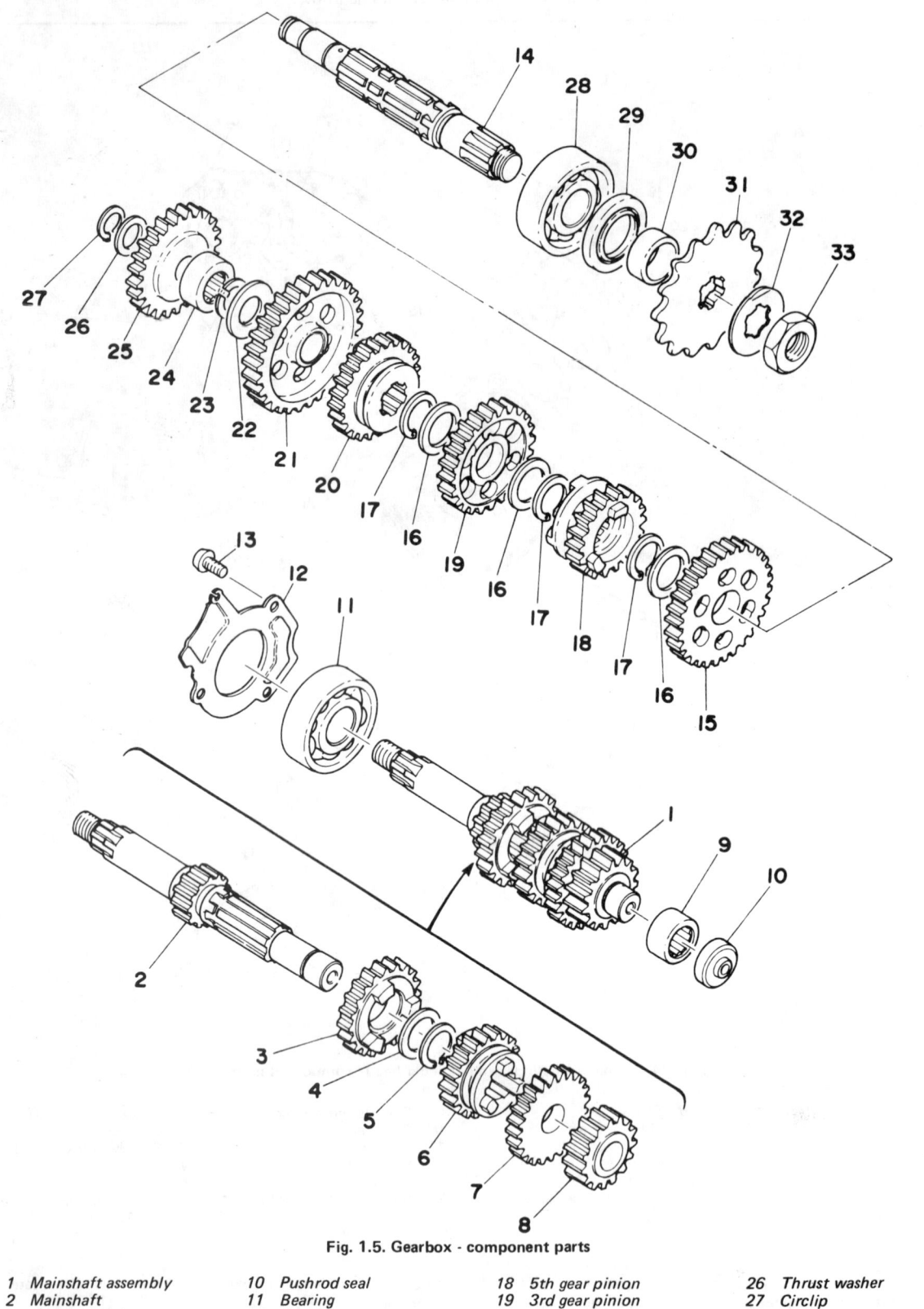

Fig. 1.5. Gearbox - component parts

| | | | | | |
|---|---|---|---|---|---|
| 1 | Mainshaft assembly | 10 | Pushrod seal | 18 | 5th gear pinion |
| 2 | Mainshaft | 11 | Bearing | 19 | 3rd gear pinion |
| 3 | 4th gear pinion | 12 | Retainer | 20 | 4th gear pinion |
| 4 | Washer | 13 | Screw - 3 off | 21 | 1st gear pinion |
| 5 | Circlip | 14 | Layshaft | 22 | Shim |
| 6 | 3rd gear pinion | 15 | 2nd gear pinion | 23 | Circlip |
| 7 | 5th gear pinion | 16 | Washer - 3 off | 24 | Bearing |
| 8 | 2nd gear pinion | 17 | Circlip 3 0ff | 25 | Kickstart idler pinion |
| 9 | Bearing | | | | |

| | |
|---|---|
| 26 | Thrust washer |
| 27 | Circlip |
| 28 | Bearing |
| 29 | Oil seal |
| 30 | Spacer |
| 31 | Gearbox sprocket |
| 32 | Lockwasher |
| 33 | Nut |

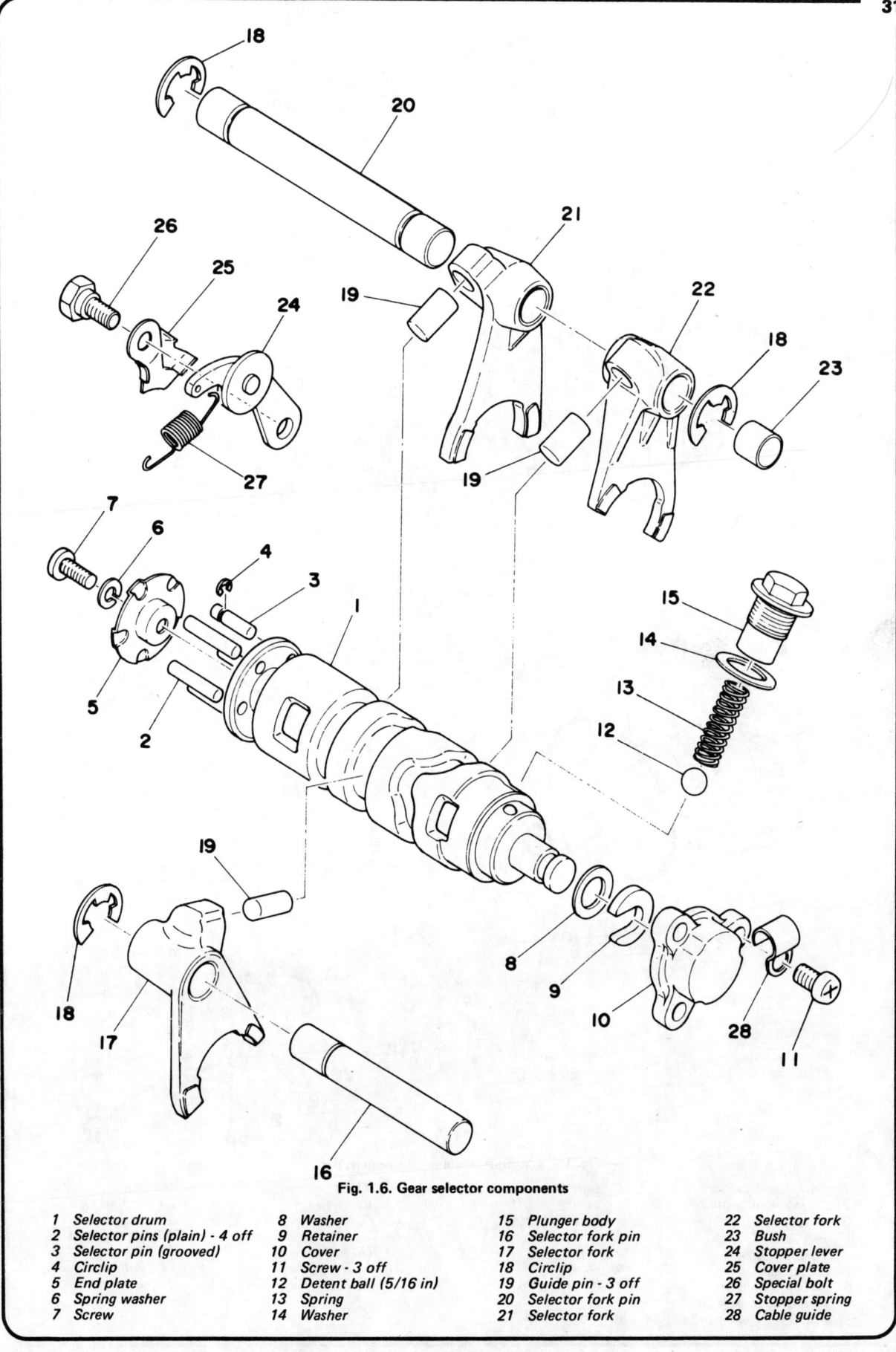

Fig. 1.6. Gear selector components

1  Selector drum
2  Selector pins (plain) - 4 off
3  Selector pin (grooved)
4  Circlip
5  End plate
6  Spring washer
7  Screw
8  Washer
9  Retainer
10 Cover
11 Screw - 3 off
12 Detent ball (5/16 in)
13 Spring
14 Washer
15 Plunger body
16 Selector fork pin
17 Selector fork
18 Circlip
19 Guide pin - 3 off
20 Selector fork pin
21 Selector fork
22 Selector fork
23 Bush
24 Stopper lever
25 Cover plate
26 Special bolt
27 Stopper spring
28 Cable guide

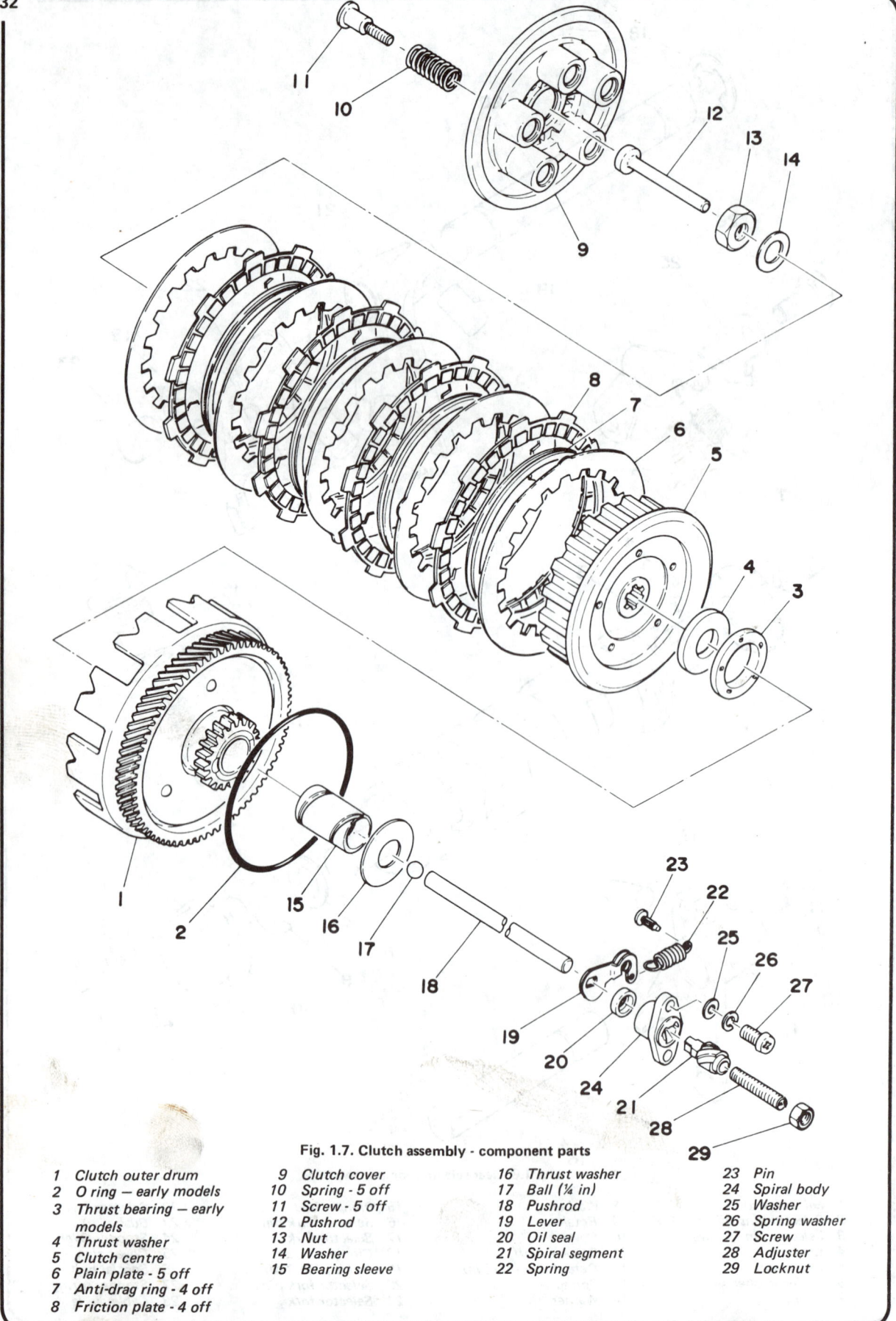

Fig. 1.7. Clutch assembly - component parts

| | | | | | |
|---|---|---|---|---|---|
| 1 | Clutch outer drum | 9 | Clutch cover | 16 | Thrust washer |
| 2 | O ring – early models | 10 | Spring - 5 off | 17 | Ball (¼ in) |
| 3 | Thrust bearing – early models | 11 | Screw - 5 off | 18 | Pushrod |
| 4 | Thrust washer | 12 | Pushrod | 19 | Lever |
| 5 | Clutch centre | 13 | Nut | 20 | Oil seal |
| 6 | Plain plate - 5 off | 14 | Washer | 21 | Spiral segment |
| 7 | Anti-drag ring - 4 off | 15 | Bearing sleeve | 22 | Spring |
| 8 | Friction plate - 4 off | | | | |
| 23 | Pin | | | | |
| 24 | Spiral body | | | | |
| 25 | Washer | | | | |
| 26 | Spring washer | | | | |
| 27 | Screw | | | | |
| 28 | Adjuster | | | | |
| 29 | Locknut | | | | |

## 23 Engine casings and covers: examination and renovation

1 The aluminium alloy casings and covers are unlikely to suffer damage through ordinary use. However, damage can occur if the machine is dropped, or if sudden mechanical mishaps occur, such as the rear chain breaking.
2 Small cracks or holes may be repaired with an epoxy resin adhesive, such as Araldite, as a temporary expedient. Permanent repairs can only be effected by argon-arc welding, and a specialist in this process is in a position to advise on the viability of the proposed repair. Often it may be cheaper to buy a new replacement.
3 Damaged threads can be reclaimed economically by using a diamond section wire insert, of the Helicoil type, which is easily fitted after drilling and re-tapping the affected thread. The process is quick and inexpensive, and does not require as much preparation and work as the older method of fitting brass, or similar inserts. Most motor cycle dealers and small engineering firms offer a service of this kind.
4 Sheared studs or screws can usually be removed with screw extractors, which consist of tapered, left-hand thread screws, of very hard steel. These are inserted by screwing anticlockwise, into a pre-drilled hole in the stud, and usually succeed in dislodging the most stubborn stud or screw. The only alternative to this is spark erosion, but as this is a very limited, specialised facility, it will probably be unavailable to most owners. It is wise, however, to consult a professional engineering firm before condemning an otherwise sound casing. Many of these firms advertise regularly in the motor cycle papers.
5 The crankshaft main bearings and gearbox bearings should be examined for wear and roughness when turned, and if suspect, should be renewed. Remove the bearing oil seal retainer and prise out the old seal, where appropriate. The bearings can be removed easily by applying heat to the casing, causing the aluminium alloy to expand at a faster rate than that of the steel bearing, allowing the bearing to become loose. The safest way of doing this is to place the casing in an oven, heating it to about 80° - 100°C. The casing can then be banged on a wooden bench or board, face down, to jar the bearing free. The new bearings can be tapped into position, using a large diameter socket as a drift. Care should always be exercised when heating alloy casings as excessive or localised heat can easily cause warpage.

## 24 Engine reassembly: general

1 Before reassembly of the engine/gear unit is commenced, the various component parts should be cleaned thoroughly and placed on a sheet of clean paper, close to the working area.
2 Make sure all traces of old gaskets have been removed and that the mating surfaces are clean and undamaged. One of the best ways to remove old gasket cement is to apply a rag soaked in methylated spirit. This acts as a solvent and will ensure that the cement is removed without resort to scraping and the consequent risk of damage.
3 Gather together all of the necessary tools and have available an oil can filled with clean engine oil. Make sure all the new gaskets and oil seals are to hand, also all the replacement parts required. Nothing is more frustrating than having to stop in the middle of a reassembly sequence because a vital gasket or replacement has been overlooked.
4 Make sure that the reassembly area is clean and that there is adequate working space. Many of the smaller bolts are easily sheared if over-tightened. Always use the correct size screwdriver bit for the crosshead screws and never an ordinary screwdriver or punch. If the existing screws show evidence of maltreatment in the past, it is advisable to renew them as a complete set.
5 If the purchase of a replacement set of screws is being contemplated, it is worthwhile considering a set of socket or Allen screws. These are invariably much more robust than the originals, and can be obtained in sets for most machines, in either black or nickel plated finishes. The manufacturers of these screw sets advertise regularly in the motor cycle press.

## 25 Engine reassembly: gear clusters and selector mechanism reassembly and replacement

1 Having examined and renewed the gearbox components as necessary, the clusters can be built up and assembled as a complete unit for installation in the engine/gearbox casings.
2 Study the line drawing carefully (Fig. 1.5) and assemble the layshaft components in the exact order shown, ensuring that the thrust washers and circlips are correctly positioned. The gearbox mainshaft (input shaft) should be tackled in a similar manner.

## 26 Engine reassembly: replacing the gearbox components

1 Lay out the assembled mainshaft and layshaft, so that the splined end of one is adjacent to the plain end of the other. Offer up the selector drum with the smaller, grooved, end towards the gearbox sprocket spline on the layshaft. Engage the selector forks in their respective grooves: The central fork runs in a groove on the 4th gear mainshaft pinion, the other two running in similar grooves in the 3rd and 5th gear layshaft pinions. Draw the assembly together to form a cluster with the relevant pinions in mesh with each other.
2 With the left-hand casing supported on blocks, offer up the cluster and lower it into position. Check that the shafts and selector drum are still in alignment Fit the two locating dowels in position in the left-hand casing.

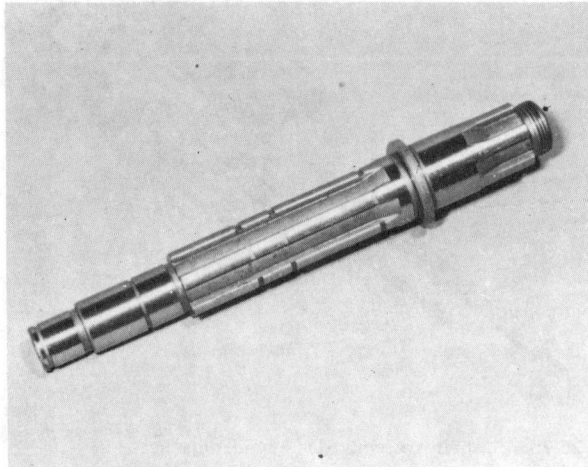

25.2a Clean and lubricate layshaft, then fit ...

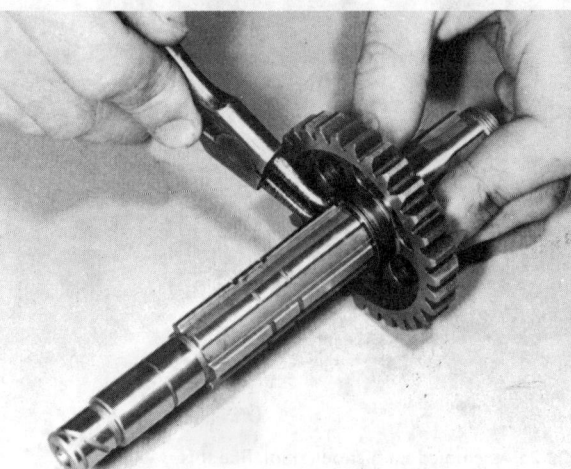

25.2b ... 2nd gear pinion, circlip and washer

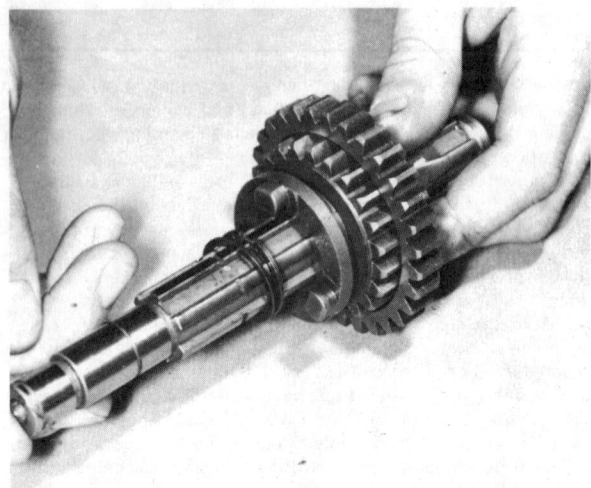

25.2c 5th gear pinion is fitted thus ...

25.2d ... followed by 3rd gear pinion

25.2e Fit 4th gear pinion, and finally ...

25.2f ... bottom gear pinion, thrust washer and circlip

25.2g Assembled shaft should look like this

25.2h Mainshaft is assembled in a similar manner

# Chapter 1: Engine, clutch and gearbox

26.1 Assemble gearbox components and fit in LH case

27.2 Fit RH crankcase half and tighten down

## 27 Preparation of crankcase jointing surfaces: joining the crankcases

1  Coat the jointing surfaces of both crankcases with a thin layer of gasket cement.
2  With the left-hand crankcase lying on its left side on the workbench, lower the right-hand crankcase on to it, taking care to locate the right-hand main bearing in its housing and the gearshafts in their respective bearings. It may be necessary to give the right-hand crankcase a few light taps with a soft faced mallet before the jointing surfaces will mate up correctly. **Do not use force.** If the crankcase will not align, one of the main bearings is not seating correctly.
3  Replace the crosshead screws in the left-hand crankcase and tighten fully.

## 28 Reassembly of the gearchange, kickstart mechanism and clutch

1  Replace the neutral indicator switch unit, and the detent ball, spring and plug, in the left-hand casing. Fit the washer and horseshoe-shaped retainer in its groove on the end of the selector drum shaft, and replace the white plastic cover and its gasket.
2  Replace the shift drum stopper lever and spring in its location in the right-hand crankcase above the right-hand end of the gear selector barrel. The lever pivots about a stopper bolt which is screwed into the crankcase and positive location is maintained by the spring which is attached between the lever and the clutch mainshaft bearing cover plate.
3  Replace the gearchange spindle and gearshift cam assembly as one unit by passing the gearchange spindle through the hole (below the clutch mainshaft) in the right-hand crankcase. The uppermost end of the lever assembly meshes with the gear selector drum adjacent to the shift drum stopper lever. The gearchange return spring, which is mounted on the gearchange spindle, is located by a peg which is screwed into the crankcase and locked by a nut. Check that this is correctly adjusted (see Fig. 1.8)
4  The gearchange spindle is restrained from endfloat or sideways movement by a plain thrust washer and circlip which fits on the spindle where it emerges from the left-hand crankcase. Slide the thrust washer over the gearchange spindle and press the circlip home in its groove with a pair of pliers.

5  Replace the kickstart idler pinion over the idler shaft which is adjacent and to the left of the clutch mainshaft and replace the washer and circlip which retains it.
6  Replace the kickstart pinion, kickstart spindle and kickstart return spring as one unit to its boss, at the rear of the right-hand crankcase.
7  Fit the mainshaft pinion, so that it engages on the Woodruff key, followed by its plain washer and nut. It is easier to tighten the nut finally when the clutch has been fitted.
8  Slide the thrust washer, which locates adjacent to the mainshaft main bearing, over the clutch mainshaft. Fit the clutch bearing sleeve.
9  Fit the clutch outer drum, and its thrust washer, over the clutch sleeve.
10  Replace the clutch centre, and secure it by tightening the centre nut. The centre can be locked by using the made-up clutch holding tool mentioned earlier or by passing a piece of 1/8 in steel strip, about an inch wide, through the slots in the drum, to engage in the grooves in the clutch centre. Care should be exercised in this latter case, as the aluminium alloy of both the drum and the centre is easily damaged. Fit the mushroom headed clutch pushrod, having first greased it lightly.
11  Refit the five plain plates, four friction plates and four anti-drag rings; fit the plates alternately as shown in the accompanying photographs and Fig. 1.7. Note that the notches on the periphery of the plain plates must be arranged equally around the clutch centre; position each notch at 72° intervals.
12  Refit the cover, springs and the five clutch retaining screws. Tighten the screws diagonally and evenly until the heads are just level with the cover.
13  Finally, tighten the crankshaft pinion nut at this stage, by wedging a small wad of rag between it and the clutch drum gear surface. Tighten to 3 - 5 kg f m (22 - 36 lb f ft).
14  Fit the white plastic oil pump and tachometer drive gear on its supporting pin, not omitting the two thrust washers (where fitted).
15  Clean the jointing surfaces of the engine/gearbox casing and the right-hand cover. Smear a thin film of gasket cement on the mating surfaces, and fit a new gasket to the engine/gearbox casing. Offer up the cover, ensuring that the plastic oil pump - tachometer pinion engages correctly. Replace the crossheaded screws and tighten in a diagonal sequence, not forgetting the two screws recessed in the oil pump well.

28.1a Fit thrust washer, followed by ...

28.1b ... U-shaped retainer

28.2 Fit the selector drum stopper and spring

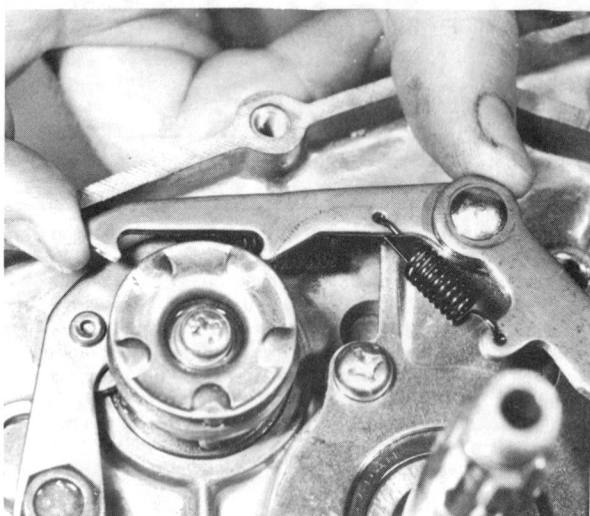
28.3a Fit selector shaft, ensuring claw, and ...

28.3b ... return spring engage correctly

28.5 Fit kickstart idler pinion, washer and circlip ...

28.6 ... and kickstart mechanism (cover removed to show spring position)

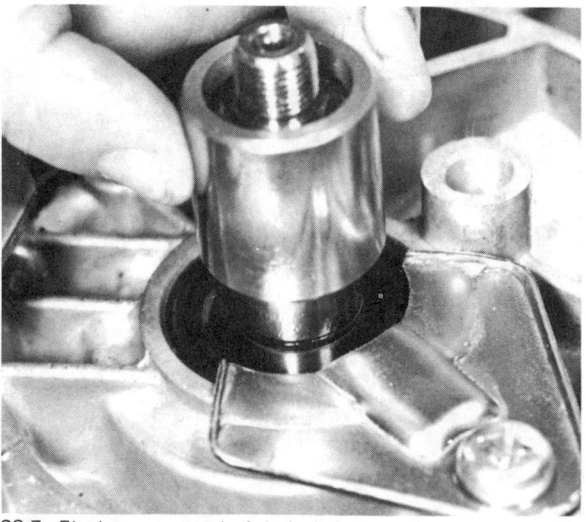

28.7a Fit sleeve over end of shaft, followed by ...

28.7b ... Woodruff key and pinion

28.7c Nut can be run onto thread - tighten later

28.8 Fit thrust washer to shaft

28.9a Clutch drum runs on ...

28.9b ... sleeve over shaft

28.9c Fit clutch drum, followed by ...

28.10a ... washer and centre

28.10b Fit washer and centre nut ...

28.10c ... and mushroom-headed pushrod

28.11a Plain plate to be installed first, followed by ...

## Chapter 1: Engine, clutch and gearbox

28.11b ... friction plate. Note position of ...

28.11c ... rubber anti-drag rings between friction plates

28.11d Finally, fit cover, springs and retaining screws

28.15 Right-hand outer cover can now be fitted

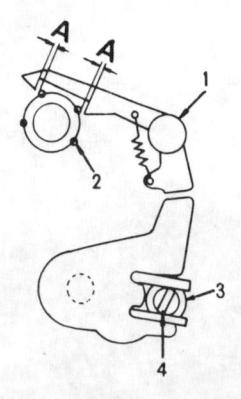

Fig. 1.8. Setting the selector arm

*A and A should be equidistant*
1. *Selector arm*
2. *Selector drum pins*
3. *Adjuster locknut*
4. *Eccentric adjuster*

### 29 Replacing the gearbox sprocket and generator

1  Fit the gearbox sprocket on its taper on the left-hand side of the engine/gearbox unit. Fit the tab washer and nut, but do not tighten it until the engine unit has been reinstalled in the frame.

2  Place the generator stator in position and fit the two countersunk screws which retain it. Ensure that the stator sits squarely against the crankcase. Place the Woodruff key in its slot in the crankshaft, and slide the generator flywheel into position. Tap it onto its taper and fit the washer and securing nut. Lock up the crankshaft in the same way as that used during the removal sequence, finally tightening the securing nut to 4 - 4.5 kg f m (29 - 33 lb f ft).

### 30 Engine reassembly: replacing the piston and rings

1  Position the engine to rest on the base of the crankcase. Pad the mouth of the crankcase with clean rag prior to fitting the piston and piston rings, so that any displaced parts will be prevented from falling in.

2  Replace the caged needle roller bearing in the small end, then

fit the piston and gudgeon pin, checking to ensure that it is replaced securely. Note that the piston has an arrow stamped on the crown, which must face forwards.
3  If the gudgeon pin is a tight fit in the piston bosses, the piston can be warmed with warm water to effect the necessary temporary expansion. Oil the gudgeon pin and piston bosses before the gudgeon pin is inserted, then fit the circlips, making sure that they are engaged fully with their retaining grooves. A good fit is essential, since a displaced circlip will cause extensive engine damage. Always fit new circlips. NEVER re-use the old ones.
4  Check that the piston rings are fitted correctly, with their ends either side of the ring pegs.
If this precaution is not observed, the rings will be broken during assembly.

## 31 Engine reassembly: replacing the cylinder barrel

1  Place a new cylinder base gasket over the retaining studs and lubricate the cylinder bore with clean engine oil. Arrange the piston so that it is at top dead centre (TDC) and lower the cylinder down the retaining studs until contact is made with the piston. The rings can now be squeezed one at the time until the cylinder barrel will slide over them, checking to ensure that the ends are still each side of the ring peg. Great care is necessary during this operation, since the rings are brittle and very easily broken.
2  Although the cylinder barrel has a good lead-in, to facilitate entry of the piston rings, a piston ring clamp can be used as an alternative to the hand feed method. Here again, care must be taken to ensure that the rings are correctly positioned in relation to the piston ring pegs.
3  When the rings have engaged fully with the cylinder bore withdraw the rag packing from the crankcase mouth and slide the cylinder barrel down the retaining studs, so that it seats on the new base gasket (no gasket cement).

## 32 Engine reassembly: replacing the cylinder head

1  Place a new copper cylinder head gasket on the top of the cylinder barrel, using a smear of grease to retain it in position. Fit the cylinder head, taking care that the cylinder head gasket is not displaced or distorted during the initial tightening down.
2  The cylinder head has four sleeve bolts, which must be tightened evenly, in a diagonal sequence. This is most important, since distortion will occur if this precaution is not observed. Use a torque wrench to achieve the final setting of 2 kg f m (15 lb f ft).

3  Replace the spark plug in order to prevent any extraneous material from dropping into the engine whilst it is being refitted into the frame.

## 33 Refitting the engine/gear unit in the frame

1  Place the machine on the centre stand, so that it is standing rigidly on firm ground. Lift the engine unit and with the aid of an assistant, slide the engine unit into the frame from the left-hand side.
2  When the engine unit is in approximately the correct position fit the front and upper rear mounting bolts loosely.
3  Replace the footrest assembly and retain it with its single short bolt before fitting the rear engine mounting bolts. Finally tighten all mounting bolts, when correctly positioned.
4  Refit the rear drive chain, the spring link being fitted with the closed end facing the direction of travel. The gearbox sprocket nut may be finally tightened by holding the rear brake on to prevent rotation. Always peen the tab washer into position to prevent the nut from becoming slack in use.

30.2a Small end bearing is sliding fit in connecting rod eye

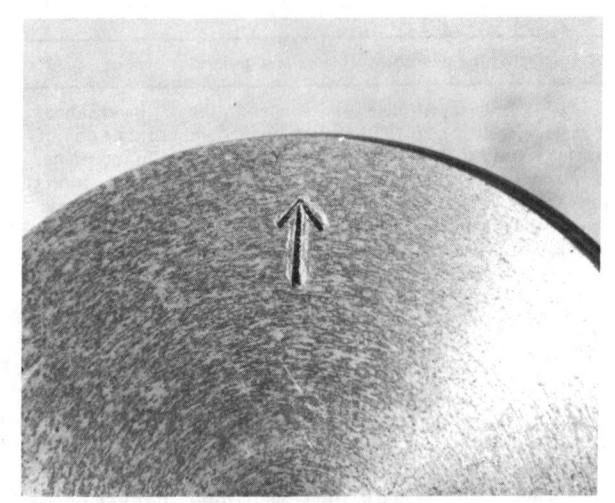

30.2b Note arrow on piston denoting front

30.3 Fit piston, having packed crankcase mouth to catch dropped clips

## 34 Engine reassembly: reconnecting the carburettor, fuel, oil and drain hoses

1  Place the carburettor back in position and replace and tighten the two socket screws which retain it to the reed-valve case, taking care not to overtighten them as this can easily warp the mounting flange. Refit the air hose between the carburettor intake and the air cleaner case. Refit the oil delivery hose to its stub on the inlet manifold.

2  Reconnect the carburettor breather and drain pipes to their respective take off points on the carburettor body, and route them down behind the engine unit. Fit the throttle valve into the carburettor body, ensuring that it locates correctly in its bore. There is a small slot and corresponding pip which act as a guide. Finally, fit and tighten the carburettor top.

## 35 Engine reassembly: reconnecting, bleeding and setting the oil pump

1  Reconnect the wire control cable linking the oil pump with the twist grip throttle. It is best to arrange the wire in a loop and engage the nipple with the oil pump pulley, before the cable is seated in the pulley groove.

2  Because the oil feed pipe now contains air, it is necessary to bleed the oil pump until all the air bubbles are removed. Check that the oil tank is filled with oil, then unscrew and remove the small crosshead screw, which has a fibre washer beneath the head, from the oil pump body. Rotate the plastic wheel with the milled edge at the rear of the oil pump, in a **clockwise** direction (as denoted by the arrow marking) and continue turning until the oil commences to flow from the outlet which the bleed screw normally seals. Continue turning until all air bubbles have been eliminated from the main feed, then replace the screw and washer.

3  To check whether the pump opening is correct, follow the procedure given in Chapter 2, Section 17. If the pump was set up correctly initially, it is improbable that significant changes in setting will be required. **Do not omit this check under the assumption it must be correct.**

4  Replace the semi-circular cover over the oil pump, which is retained by three crosshead screws.

## 36 Engine reassembly: replacing the left-hand outer casing

1  Refit the clutch cable to the operating arm which is attached to the inside of the cover.

2  Grease the clutch pushrod and insert the ball, followed by the pushrod, into position in the gearbox mainshaft. Make sure that both locating dowels are fitted, before offering up the cover and fitting the retaining screws. Tighten progressively in a diagonal sequence to avoid warping the cover.

3  Slacken off the clutch cable at the handlebar lever adjuster. Slacken off the locknut, and screw the adjuster screw in until all play in the actuating mechamism is removed, then back it off by ¼ turn and tighten the locknut. Adjust cable free play at the handlebar lever, to give 2 - 3 mm (0.07 - 0.12 in) free play between the lever end and its mounting bracket.

## 37 Engine reassembly: replacing the exhaust pipe and silencer

1  The silencer is retained to the frame by two mounting points, one being a nut and stud which attaches directly to the frame lug, and the other a bracket, retained by the lower rear engine bolt. Make sure that the engine bolt is fully tightened after the silencer is fitted.

2  The exhaust pipe is retained by a flange and two studs to the cylinder. The other end pushes into the silencer, and is clamped in position by a large gland nut. Always tighten the flange nuts first, so that the pipe assumes its natural position, before tightening the gland nuts on the silencer.

33.3 Lift assembled engine unit into position in frame

34.1 Drain and breather tubes must be connected - Note plugged oil pipe

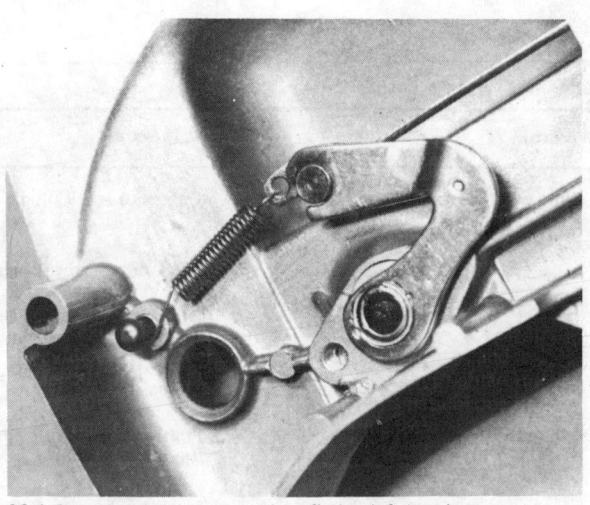

36.1 Connect clutch cable before fitting left-hand cover

## 38 Engine reassembly: final adjustments

1 Reconnect the wiring block between the generator and harness. This connector pushes together and is not reversable. Replace the neutral switch lead and fit the HT lead to the spark plug. Check that the ignition and lighting systems are functioning correctly when switched on.
2 Refit the kickstart pedal, ensuring that it will not strike and damage the casing. Replace and tighten the pinch bolt. The gearchange lever is retained by a pinch bolt where it joins the splined shaft. Check that the pedal is in the most convenient position before finally tightening the pinch bolt.
3 Replace the petrol tank, making sure that the rubber buffers engage in the cups on the steering head. Fit the rear lug over its rubber pad. Reconnect the pipe joining the two tank halves, and the pipe between the fuel tap and carburettor.
4 Check and if necessary, top up the oil tank. Remove the gearbox filler cap from the right-hand cover and add 700 cc of engine oil (SAE 10W/30) to the gearbox assembly.

## 39 Starting and running the rebuilt engine

1 When the initial start-up is made, run the engine slowly for the first few minutes, especially if the engine has been rebored or a new crankshaft fitted. Check that all the controls function correctly and that there are no oil leaks, before taking the machine on the road. The exhaust will emit a high proportion of white smoke during the first few miles, as the excess oil used whilst the engine was reassembled is burnt away. The volume of smoke should gradually diminish until only the customary light blue haze is observed during normal running. It is wise to carry a spare spark plug during the first run, since the existing plug may oil up due to the temporary excess of oil.
2 Remember that a good seal between the piston and the cylinder barrel is essential for the correct functioning of the engine. A rebored two-stroke engine will require more careful running-in, over a longer period, than its four-stroke counterpart. There is a far greater risk of engine seizure during the first hundred miles if the engine is permitted to work hard.
3 Do not tamper with the exhaust system or run the engine without baffles fitted to the silencer. Unwarranted changes in the exhaust system will have a very marked effect on engine performance, invariably for the worse. The same applies when dispensing with the air cleaner or the air cleaner element.
4 Do not on any account add oil to the petrol under the mistaken belief that a little extra oil will improve the engine lubrication. Apart from creating excess smoke, the addition of oil will make the mixture much weaker, with the consequent risk of overheating and engine seizure. The oil pump alone should provide full engine lubrication.

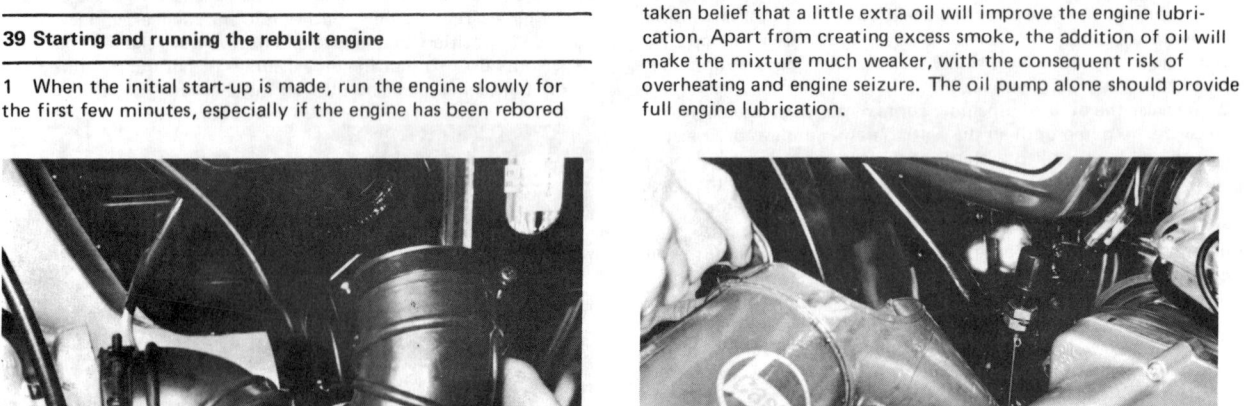

38.1 Reconnect air trunking between filter and carburettor

38.4 Fill gearbox to level on dipstick

## 40 Fault diagnosis: Engine

| Symptom | Cause | Remedy |
| --- | --- | --- |
| Engine will not start | Defective spark plug | Remove plug and lay on cylinder head. Check whether spark occurs when engine is kicked over. |
| | Dirty or closed contact breaker points | Check condition of points, and whether gap is correct. |
| | Discharged battery | Check whether lights work. If battery is flat, remove and charge. |
| | Air leak at crankcase or worn crankshaft oil seals | Flood carburettor and check whether petrol is reaching the plug. |
| Engine runs unevenly | Ignition and/or fuel system fault | Check as though engine will not start. |
| | Blowing cylinder head gasket | Oil leak should provide evidence. Renew gasket. |
| | Incorrect ignition timing | Check and if necessary adjust. |
| | Choked silencer | Remove baffles and clean. |

## Chapter 1: Engine, clutch and gearbox

| | | |
|---|---|---|
| Lack of power | Incorrect ignition timing | See above. |
| | Fault in fuel system. | Check system and vent in filler cap. |
| | Choked silencer | See above. |
| White smoke from exhaust | Too much oil | Check oil pump setting. |
| | Engine needs rebore | Rebore and fit oversize piston. |
| | Tank contains two-stroke petroil and not straight petrol. | Drain and refill with straight petrol. |
| Engine overheats | Pre-ignition and/or weak mixture | Check carburettor settings, also grade of plug fitted. |
| | Lubrication failure | Stop engine and check oil pump setting. Is oil tank dry? |

### 41 Fault diagnosis: Clutch

| Symptom | Cause | Remedy |
|---|---|---|
| Engine speed increases but machine does not respond | Clutch slip | Check whether clutch adjustment still has free play. Check thickness of linings and renew if near wear limit. |
| Difficulty in engaging gears, gear changes jerky and machine creeps forward, even when clutch is withdrawn fully | Clutch drag | Check clutch adjustment to eliminate excess play. Check whether clutch centre and outer drum have indented slots. |
| Operating action stiff | Clutch assembly loose on mainshaft | Check tightness of retaining nut. |
| | Bent pushrod | Renew. |
| | Dry pushrod | Lubricate. |
| | Damaged, trapped or frayed control cable | Check cable and renew if necessary. Make sure cable is lubricated and has no sharp bends. |

### 42 Fault diagnosis: Gearbox

| Symptom | Cause | Remedy |
|---|---|---|
| Difficulty in engaging gears | Selector forks or rods bent | Renew. |
| | Broken springs in gear selector mechanism | Check and renew. |
| | Clutch drag | See above Section. |
| Machine jumps out of gear | Worn dogs on ends of gear pinions | Strip gearbox and renew worn parts. |
| | Sticking detent plunger | Remove plunger cap and free plunger assembly. |
| Kickstart does not return | Broken return spring | Remove right-hand crankcase cover and renew spring. |
| Kickstart slips or jams | Worn ratchet assembly | Remove right-hand crankcase cover, dismantle kickstart assembly and renew worn parts. |
| Gear change lever does not return | Broken return spring | Remove right-hand crankcase cover and renew spring. |

# Chapter 2 Fuel system and lubrication

*For information relating to 1977 on models refer to Chapter 7*

## Contents

| | |
|---|---|
| General description ... 1 | Air cleaner: removing, cleaning and replacing the element ... 11 |
| Petrol tank: removal and replacement ... 2 | Crankcase drain plug ... 12 |
| Petrol tap: removal, dismantling and replacement ... 3 | Exhaust pipe and silencer: examining and cleaning ... 13 |
| Petrol feed pipe: examination ... 4 | The lubrication system ... 14 |
| Carburettor: removal ... 5 | Removing and replacing the oil pump ... 15 |
| Carburettor: dismantling and examination ... 6 | Bleeding the oil pump ... 16 |
| Carburettor: reassembly ... 7 | Checking the oil pump setting ... 17 |
| Carburettor: checking the settings ... 8 | Removing and replacing the oil tank ... 18 |
| Reed valve induction system: mode of operation ... 9 | Fault diagnosis: fuel system ... 19 |
| Reed valve: removal, examination and renovation ... 10 | Fault diagnosis: lubrication system ... 20 |

## Specifications

**Fuel tank capacity**
- Early type ... 9 litres (1.97 Imp gall) (2.38 US gall)
- Late type ... 10 litres (2.20 Imp gall) (2.64 US gall)

**Oil tank capacity**
- Early type ... 1.5 litres (2.63 Imp pints) (1.6 US quarts)
- Late type ... 1.3 litres (2.3 Imp pints) (1.4 US quarts)

**Carburettor**

| | RS100 | RS125 |
|---|---|---|
| Make ... | Mikuni | Mikuni |
| Type ... | *VM 20 SH (IY800) | VM 24 SH (2A000) |
| Main jet | 125 | 125 |
| Air jet | 2.5 | 0.5 |
| Jet needle | *4J21 | 512 |
| Clip position | *2 | 2 |
| Needle jet | *P - 2 | E - 4 |
| Throttle valve | 1.5 | 2.0 |
| Pilot jet | 20 | 30 |
| Air screw setting | 1¾ turns out | 1¼ turns out |
| Starting jet | 30 | 40 |
| Fuel level | 26 ± 1.0 mm (1.02 ± 0.04 in) | |
| Float height | 21 ± 1.0 mm (0.83 ± 0.04 in) | |
| Normal idling speed | 1300 ± 50 rpm | 1200 ± 50 rpm |

*Note: On RS100 models fitted with a VM 20 SH Mikuni carburettor having a IY890 suffix the following settings apply:

- Jet needle ... 4F22
- Clip position ... 4
- Needle jet ... 0.6

**Oil pump**

| | RS100 | RS125 |
|---|---|---|
| Colour code ... | Orange | Dark blue |
| Output per 200 strokes: | | |
| Minimum ... | 0.5 - 0.63 cc | |
| Maximum ... | 4.65 - 5.15 cc | |
| Minimum stroke ... | 0.20 - 0.25 mm (0.008 - 0.040 in) | |
| Maximum stroke ... | 1.85 - 2.05 mm (0.073 - 0.081 in) | |

# Chapter 2: Fuel system and lubrication

## 1 General description

The fuel system comprises a petrol tank, from which petrol is gravity fed, via a petrol tap and filter unit, to the float chamber of the Mikuni carburettor.

For cold starting, the carburettor is fitted with a hand operated choke control. This device provides an enriched mixture, necessary for starting in cold conditions, and should be returned to its off position as soon as the engine has warmed up enough to accept the normal running mixture.

The lubrication is effected by a separate oil pump and feed system, similar to that pioneered by a British manufacturer in the 1930s.

The Yamaha Autolube system consists of an engine driven plunger oil pump ......... within the front of the right-hand outer ......... oil tank to the pump. ......... the engine speed, and ......... twistgrip which varies the ......... hat the correct amount of ......... engine speeds and loads, and ......... mixed two-stroke fuel are

## 2 Petrol tank: removal and replacement

1  Although it is not necessary to remove the petrol tank when the engine unit is removed from the frame, better access is gained and there is less risk of damage to the painted surface if the tank is out of the way. Apart from occasions such as these, there is rarely any need to remove the tank unless rust has formed inside as the result of long storage or if it needs repainting.

2  The petrol tank is not rigidly attached to the frame at any point, being supported and retained by two rubber buffers at the front, and the dualseat at the rear. The front rubber buffers engage in cups each side of the headstock. Before removing the tank, it must be drained, the pipe connecting the tank halves detached at one end, and the securing bolt slackened.

3  Lift the rear of the tank slightly, and pull gently rearwards to disengage the rubber buffers. The tank can be lifted away.

4  When replacing the tank, check that the mounting rubbers locate correctly, that they are in good condition, and that the joining pipe is refitted correctly and shows no sign of leaking.

## 3 Petrol tap: removal, dismantling and replacement

1  The petrol tap is secured to the left-hand underside of the petrol tank by a thread which screws into the tank base, and a locknut. The filter bowl threads into the main body of the petrol tap and can be removed by applying a spanner to the hexagon in the base of the bowl. There is a sealing washer between the filter bowl and the main body of the petrol tap to preserve a petrol-tight joint; the filter gauze is located above the filter bowl, attached to the main body of the petrol tap by a small crossheaded screw.

2  There is seldom need to disturb the main body of the petrol tap. In the event of a leak at the operating lever, the complete lever assembly can be dismantled (provided the petrol tank is drained first) with the main body undisturbed. Remove the two crosshead screws that retain the tap lever plate in position and withdraw the lever complete with plate, crinkle washer and the valve insert behind the lever. The valve is the item to be replaced if leakage occurs; it is composed of a synthetic rubber material which may commence to disintegrate after an extended period of service.

3  Before reassembling the petrol tap, check that all parts are clean, especially the two tubes (short tube reserve, long tube main feed) which extend into the petrol tank, the filter and filter body assembly. A new gasket should be fitted to the filter bowl assembly in order to effect a satisfactory seal.

4  Do not overtighten any of the petrol tap components during reassembly. The castings are in a zinc-based alloy, which will fracture easily if over-stressed. Most leakages occur as the result of defective seal.

## 4 Petrol feed pipe: examination

The petrol feed pipe, connecting the carburettor to the fuel tap, is made of thin-walled synthetic rubber, and is retained by small wire clips. Check that the pipe has not split or become brittle, due to age and the effects of heat and fuel. Check also the various drain and breather pipes.

## 5 Carburettor: removal

1  The Mikuni carburettor is retained by a flange mounting to the reed-valve case. Before it can be detached, it is first necessary to unscrew the carburettor top and withdraw the throttle valve assembly complete with the needle and return spring. These components should be left attached to the throttle cable and placed out of harm's way while the rest of the instrument is removed.

2  Make sure that the fuel tap is in the off position, then prise off the petrol feed pipe after displacing the retaining clip. Slacken and remove the two socket screws which retain the carburettor to the reed-valve case and lift the instrument away, taking care not to damage the 'O' ring. Note that it is possible to remove the carburettor and the reed-valve as a unit, if desired, but there is nothing to be gained by doing this, and the reed-valve is normally best left undisturbed.

## 6 Carburettor: dismantling and examination

1  Before commencing carburettor dismantling, the outside of the instrument should be cleaned with petrol to remove the inevitable accumulations of road grime. Lay out some sheets of clean paper to place the carburettor parts on, and keep everything as clean as possible.

The float chamber is retained to the carburettor body by four screws. Remove the screws and lift the float chamber away. The gasket should preferably be discarded and renewed, to maintain a petrol-tight joint.

2  Push out the float pivot pin which holds the float to the two cast lugs which form its attachment point. The pin is not a tight fit and can be shaken out or pushed out with a piece of wire. Lift the float unit away, taking care not to lose the pivot pin.

3  If symptoms of flooding have been in evidence, check that the float is not leaking, by shaking and listening for petrol inside. It is rare to find leaks in plastic floats, this problem being more common in the brass type.

4  A more likely cause of flooding is dirt on the float needle or its seat. The needle can easily be removed, with the float detached, by inverting the body and allowing it to drop from its seating. Examine the faces of the needle and seat for foreign matter and also for scoring. If in bad condition, renew, preferably as a pair. The needle seat can be removed using a box spanner or thin walled socket.

5  The main jet screws into the base of the banjo bolt which is screwed into the float blow. It is not prone to any real degree of wear, but can become blocked by contaminants in the petrol. These can be cleared by an air jet, either from an air line or a footpump. As a last resort, a fine bristle from a nailbrush or similar may be used, but on no account should wire be used as this may damage the precision drilling of the jet.

6  The needle jet may become worn after a considerable mileage has been covered and should be renewed along with the needle. It is screwed into the centre of the carburettor body.

7  The pilot jet is located adjacent to the needle jet, and is easily removed for cleaning.

8  Examine the throttle valve for scoring or wear, renewing if badly badly damaged. If damage is evident, check the internal bore of

the carburettor, and if necessary renew this also. The throttle valve, spring and needle are retained to the carburettor top as an assembly, by the throttle cable.

9  Pull the return spring upwards. The cable can now be removed by sliding it from its slot, and the needle, seat and clip can be lifted clear of the valve. Note the position of the clip on the needle grooves. Check that the needle is not bent, by rolling it on a flat surface.

## 7  Carburettor: reassembly

1  Reassembly should be tackled in the reverse order of that given for dismantling, bearing the following points in mind.
2  Always ensure that each component is scrupulously clean before starting reassembly. Never force parts together. The die-cast body of the carburettor is very prone to cracking. Do not over-tighten jets, they are of brass, and the threads are easily stripped.
3  When refitting the valve assembly, lubricate it with light machine oil, and ensure that the locating pip aligns correctly with the slot in the valve.
4  Before use, check for leaks, and check that the settings are as described in the following Section. If dismantling was necessary because of contaminated fuel, remove and flush the fuel tank, and clean the filter in the tap. Water can cause persistent and erratic running faults, and may form by condensation in the tank.

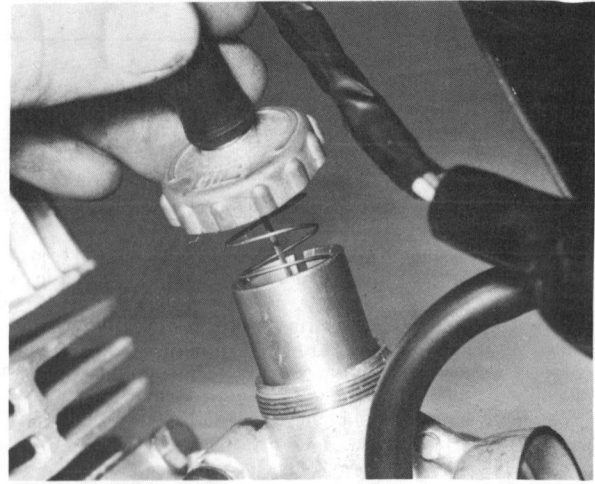

5.1 Unscrew carburettor top and withdraw valve assembly

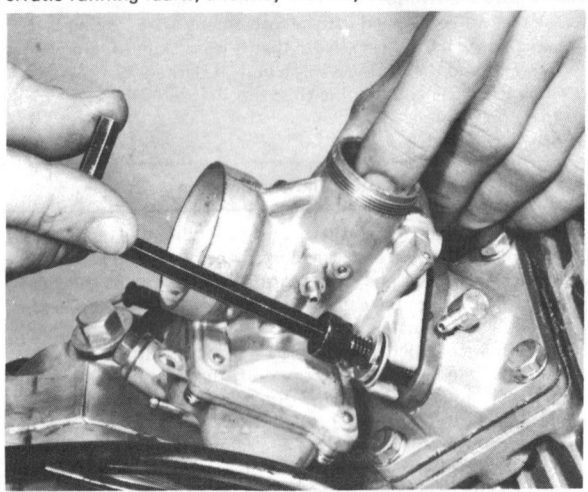

5.2a Carburettor is retained by two screws

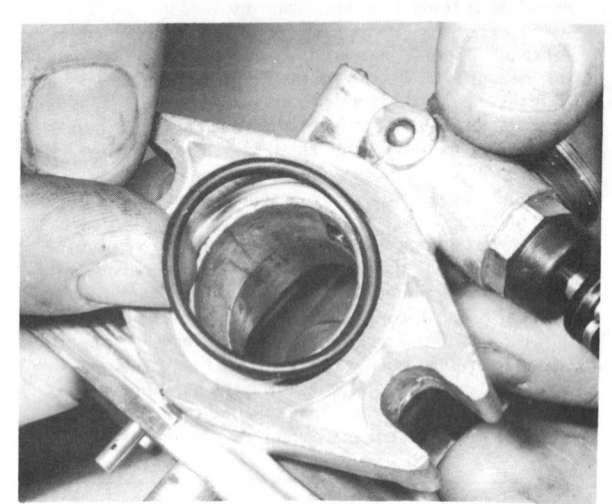

5.2b Try not to damage O ring during removal

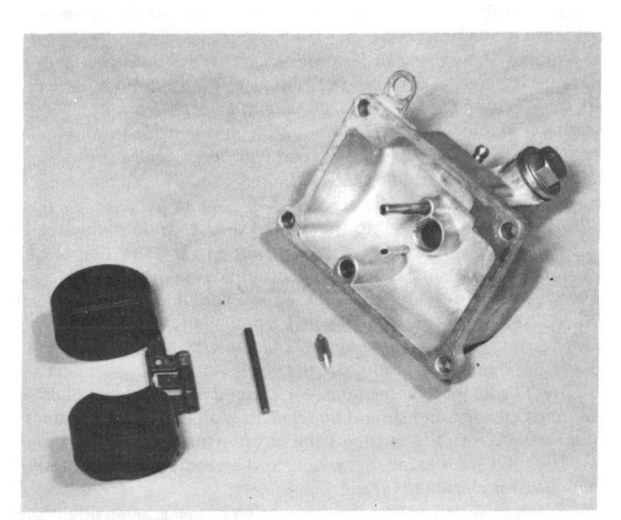

6.2a Float and float bowl components

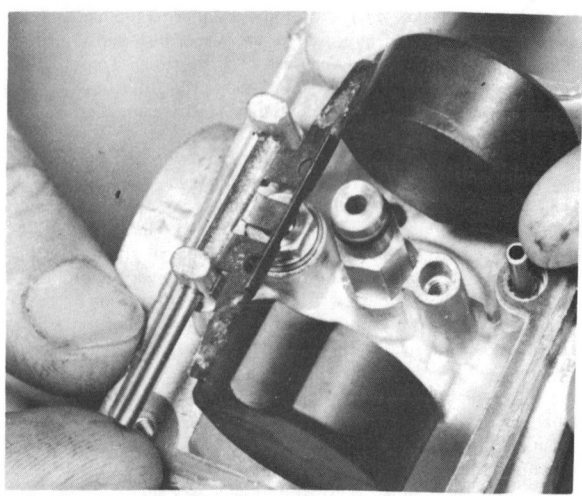

6.2b Float is retained by pivot pin

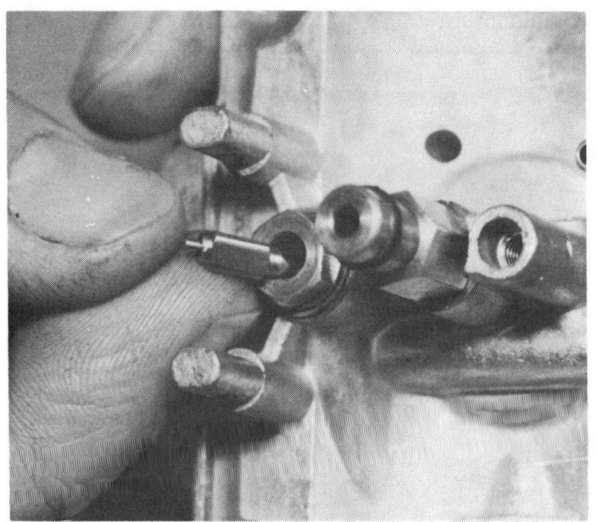

6.4a Do not lose float needle!

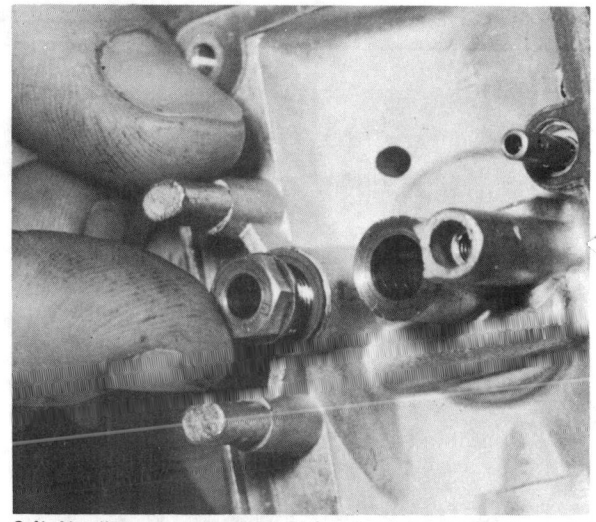

6.4b Needle seat can be removed for cleaning or renewal

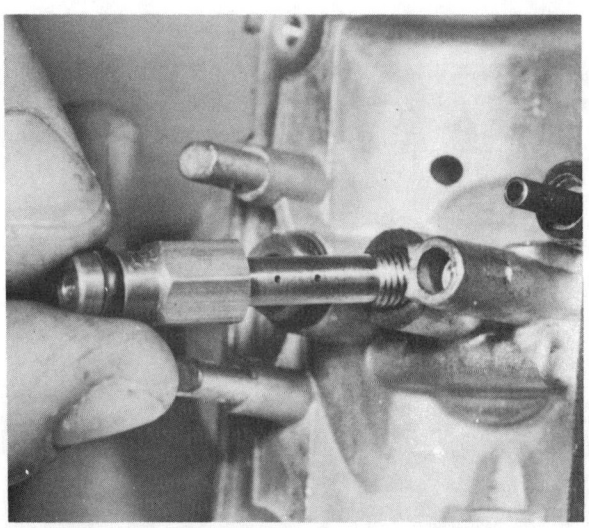

6.5 Needle jet is central in body

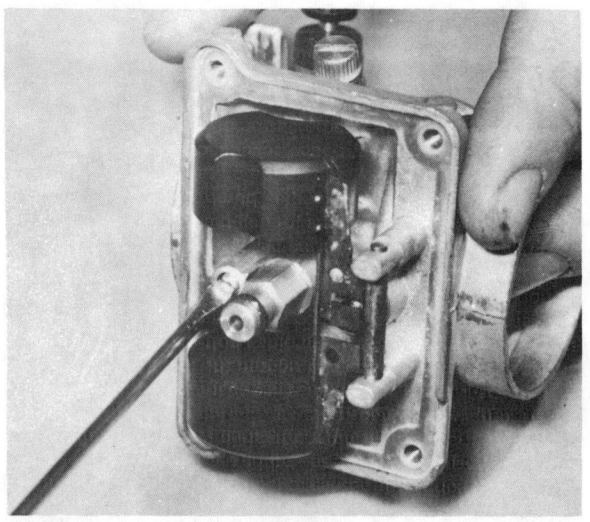

6.7 Pilot jet is adjacent to needle jet

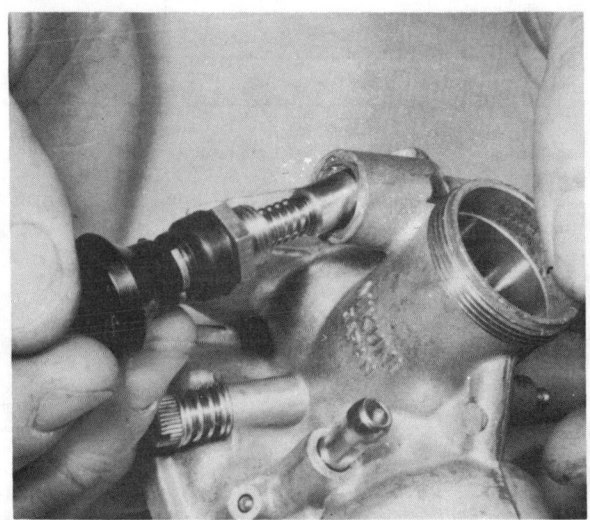

6.8a Plunger-type choke mechanism can be removed ...

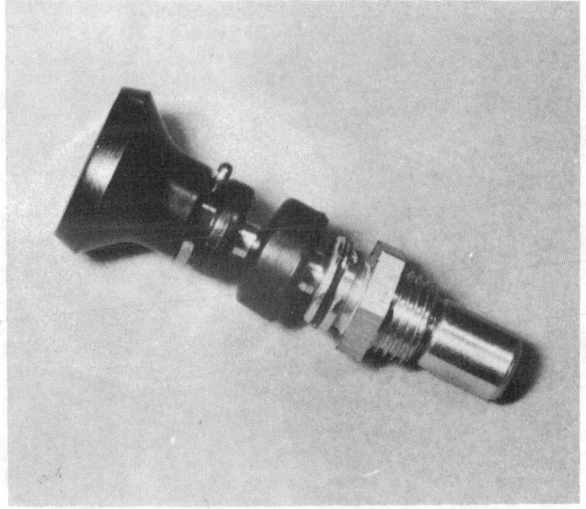

6.8b ... but will rarely require attention

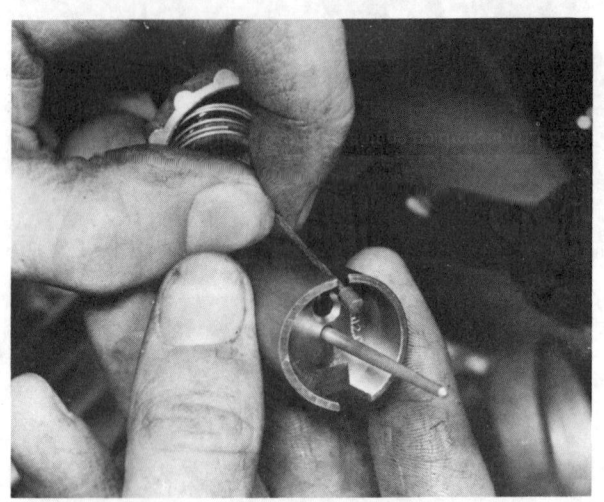

6.9a Compress spring to disengage cable from valve

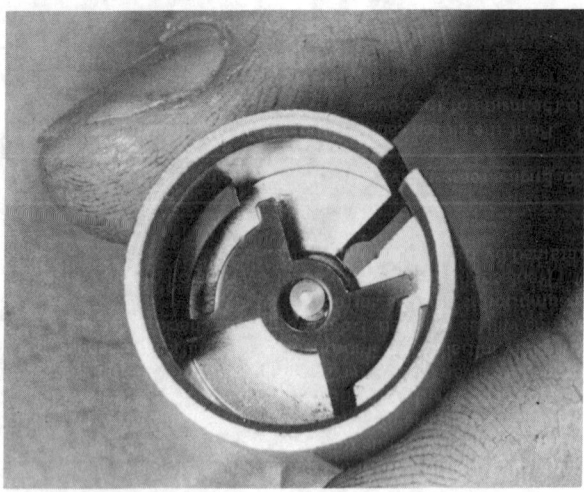

6.9b Spring seat covers needle clip, which ...

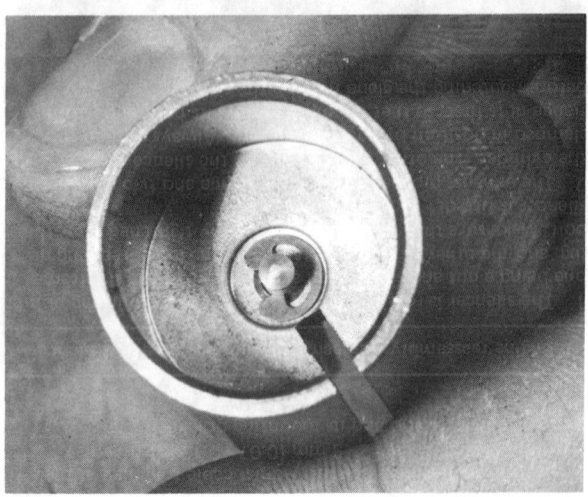

6.9c ... locates in small recess

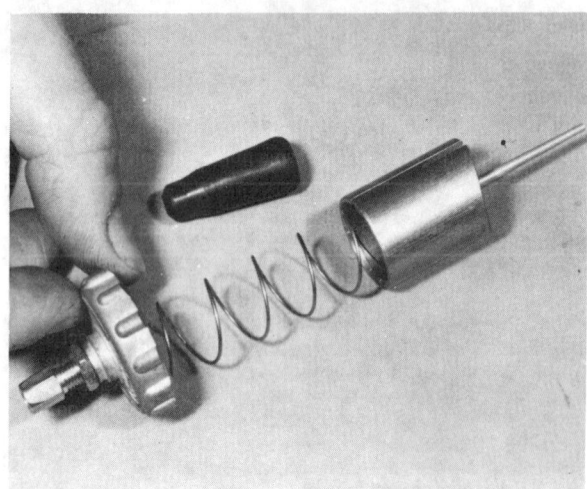

6.9d The throttle valve and top assembly

6.9e Note clip position on needle (see specifications)

## 8 Carburettor: checking the setting

1 The various jet sizes, throttle valve cutaway and needle position are predetermined by the manufacturer and should not require modification. Check with the Specifications list at the beginning of this Chapter if there is any doubt about the types fitted.

2 Slow running is controlled by a combination of the thrott' stop and pilot jet settings. Adjustment should be carried out explained in the following Section. Remember that the char eristics of the two-stroke engine are such that it is extremely difficult to obtain a slow, reliable tick-over at low rpm. If desired, there is no objection to arranging the throttle stop so that the engine will shut off completely when the throttle is closed. Unlike a petrol-lubricated engine, the oil used for engi lubrication is injected into the inlet passage of the cylinder barrel, behind the closed throttle slide. In consequence there is no risk of the engine 'drying up' when the machine is coasted down a long incline, if the throttle is closed.

3 As a rough guide, up to 1/8th throttle is controlled by the pilot jet, 1/8th to ¼ by the throttle valve cutaway, ¼ to ¾ throttle by the needle position and from ¾ to full by the size of

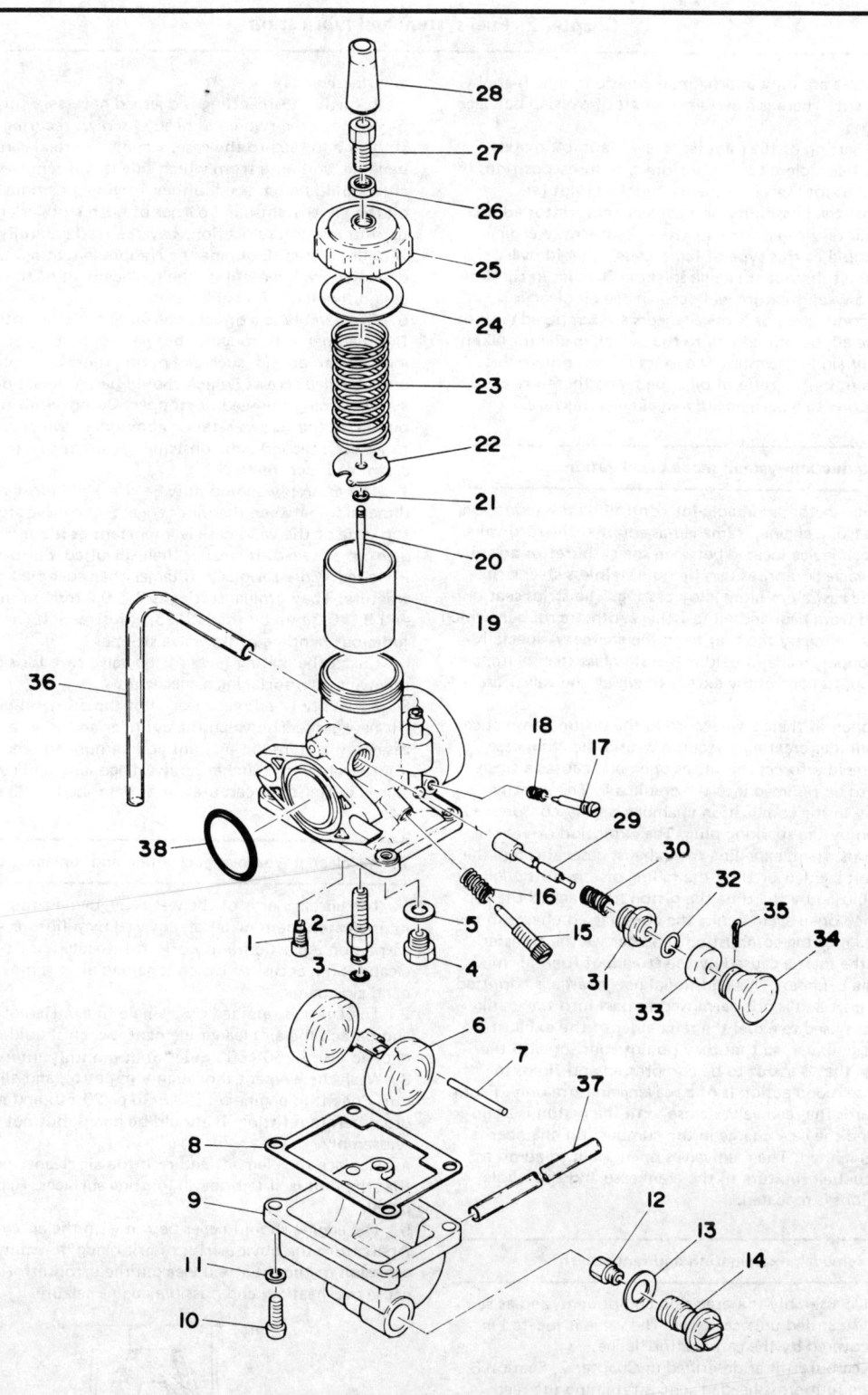

Fig. 2.1. Carburettor - component parts

| | | | |
|---|---|---|---|
| 1 Pilot jet | 11 Washer - 4 off | 21 Clip | 30 Spring |
| 2 Needle jet | 12 Main jet | 22 Retainer | 31 Body |
| 3 O ring | 13 Washer | 23 Spring | 32 Clip |
| 4 Float needle seat | 14 Banjo bolt | 24 Gasket | 33 Seal |
| 5 Washer | 15 Throttle stop screw | 25 Carburettor top | 34 Knob |
| 6 Float | 16 Spring | 26 Locknut | 35 Split pin |
| 7 Pivot pin | 17 Pilot screw | 27 Adjuster | 36 Vent pipe |
| 8 Gasket | 18 Spring | 28 Dust cover | 37 Overflow pipe |
| 9 Float bowl | 19 Throttle valve | 29 Plunger | 38 O ring |
| 10 Screw - 4 off | 20 Needle | | |

the main jet. These are only approximate divisions, which are by no means clear cut. There is a certain amount of overlap between the various stages.

4  The normal setting of the pilot jet screw is approximately one and one half complete turns out from the closed position. If the engine 'dies' at low speed, suspect a blocked pilot jet.

5  Guard against the possibility of incorrect carburettor adjustments which will result in a weak mixture. Two-stroke engines are very susceptible to this type of fault, causing rapid overheating and often subsequent engine seizure. Changes in carburation leading to a weak mixture will occur if the air cleaner is removed or disconnected, or if the silencers are tampered with in any way. Above all, do not add oil to the petrol, in the mistaken belief that it will aid lubrication. The extra oil will only reduce the petrol content by the ratio of oil added, and therefore cause the engine to run with a permanently weakened mixture.

## 9  Reed valve induction system: mode of operation

1  Of the various methods suitable for controlling the induction cycle of a two-stroke engine, Yamaha has adopted the reed valve, a non-mechanical device located between the carburettor and the cylinder. Each valve comprises two flexible stainless steel strips attached to a die-cast aluminium alloy casting. The strips seat on a gasket formed from heat and oil-resisting synthetic rubber, which is welded to the casing by the heat from the engine. A specially-shaped valve stopper, made of cold rolled stainless steel plate, acts as a form of stop, to control the extent to which the valves are free to move.

2  The valves open of their own accord as the piston commences to rise in the cylinder creating a vacuum within the crankcase. Atmospheric pressure forces the valves open and causes a fresh fuel/air charge to be rammed into the crankcase. The existing mixture, already in the combustion chamber is fully compressed and then ignited by the sparking plug. The explosion drives the piston downwards again, expelling the exhaust gases through the exhaust port as it is uncovered by the falling piston. Although the reed valves are closed by the time the piston has reached the top of its stroke, they open again whilst the compressed charge in the crankcase passes into the combustion chamber via the transfer ports, through the inertia caused by the stream of fuel/air mixture entering the cylinder. This additional new charge is admitted by the seventh port and is not permitted to pass into the crankcase. Instead, it is used to expel the remainder of the exhaust gases out of the cylinder, so that they do not conflict with the incoming charge that is about to be compressed and fired. In other words this second action is of a scavenge nature only. From this point onwards the reed valves close until the piston is again on the ascent and the new charge in the combustion chamber is compressed and ignited. The reed-valves open again to admit a new charge of fuel/air mixture to the crankcase and the whole cycle of operations is repeated.

## 10  Reed valve: removal, examination and renovation

1  The reed valve assembly is a precision component, and as such should not be dismantled unnecessarily. The valve is located in the inlet tract, covered by the carburettor flange.

2  Remove the carburettor as described in Chapter 2, Section 5, thus exposing the four cross-headed screws retaining the reed valve assembly to the cylinder. After removing these screws, the assembly can be carefully lifted away.

3  The valve can now be washed in clean petrol to facilitate further examination. It should be handled with great care, and on no account dropped. The stainless steel reeds should be inspected for signs of cracking or fatigue, and if suspect, should be renewed. Remember that any part of the assembly which breaks off in service will almost certainly be drawn into the engine, causing extensive damage.

4  Note the position of the reed in relation to the Neoprene bonded gasket against which it seats. It should be flush to form an effective seal.

If further dismantling is deemed necessary proceed as follows:
5  Remove the two cross-headed screws securing the valve stopper and reed to the case. Handle the reed carefully, avoiding bending, and note from which side it was removed. All components should be replaced in their original positions. A small cutout on the lower right-hand corner of both stopper and reed can be used to assist in relocation. Lay the reed carefully to one side if it is to be re-used. Examine the Neoprene seating face which, if defective, will necessitate the replacement of the complete alloy case, to which it is heat-bonded.

6  Reassembly is a direct reversal of the dismantling process. Clean all parts thoroughly, but gently, before refitting. A thread locking compound, such as Loctite, must be applied to the two cross-headed screws, which should be tightened progressively to avoid warping the reed or stopper. Do not omit the locking compound, as the screws retain a component which vibrates many times each second and consequently are prone to loosening if assembled incorrectly.

7  The assembly should now be checked before refitting. The dimension between the inner edge of the valve stopper and the top edge of the valve case is important as it controls the movement of the reed. If smaller than specified, performance will be impaired. More seriously, if larger than specified, the reed may fracture. The nominal setting is $7 \pm 0.3$ mm on the RS100 model and $9 \pm 0.3$ mm on the RS125, adjustment being effected by judicious bending of the valve stopper.

Check the mating faces of the valve case for signs of warping, renewing or resurfacing as necessary.

8  Assembly is a direct reversal of the dismantling sequence. Fit new gaskets between the cylinder and valve assembly, valve assembly and insulating joint and carburettor and insulating joint. Tighten the socket screws diagonally and evenly, to avoid warping. Refit the carburettor as described in Chapter 1, Section 34.

## 11  Air cleaner: removing, cleaning and replacing the element

1  The air cleaner is of the wet type, being composed of an oil-impregnated foam material, covered by a fibrous skin and carried on a plastic framework. It is located in a plastic air cleaner box, access to which is gained after removing the right-hand side panel.

2  The cover is retained by a single screw. Remove the screw and detach the air cleaner element, which should be cleaned and re-oiled every 250 - 500 miles or at monthly intervals.

3  Wash the element thoroughly in petrol, and allow it to dry out. Soak it in engine oil (SAE 30 or 20/50) and allow it to drain before refitting. It should be damp, but not dripping, on reassembly.

4  Replace the element and refit the air cleaner box lid, ensuring that the unit is airtight at all jointing surfaces. Replace the side panel.

5  The engine should never be run with the air cleaner removed. Apart from the obvious risk of grit being drawn into the engine, omission of the filter will weaken the carburettor mixture and cause overheating, and possible engine seizure.

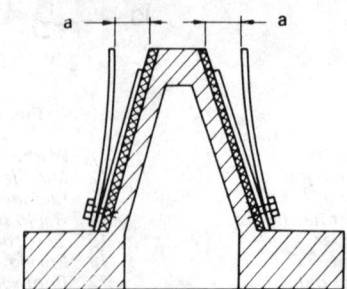

**Fig. 2.2. Reed valve stopper clearance**
RS100   RS125
Clearance at 'a': $7 \pm 0.3$ mm    $9 \pm 0.3$ mm

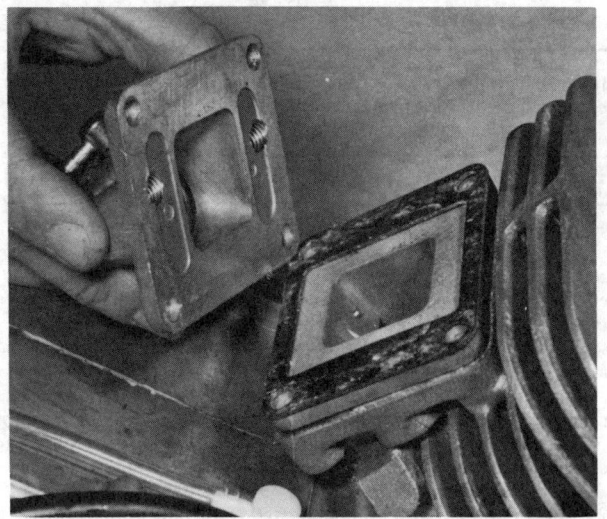

10.2a Mounting block is retained by four bolts

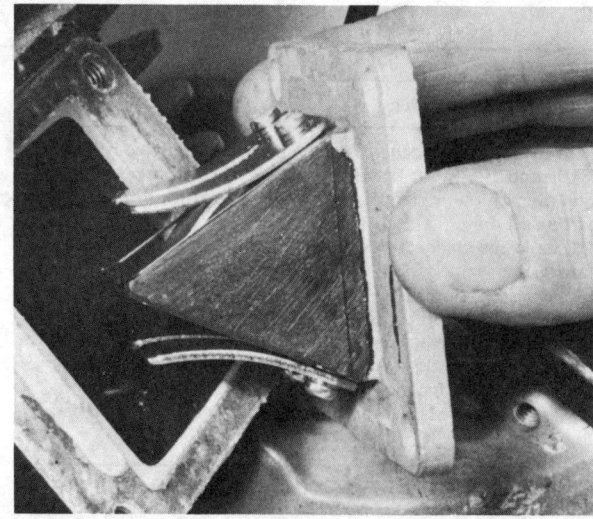

10.2b Reed valve unit can be lifted out of port

10.4 The reed valve unit complete

10.8 Reeds in closed position

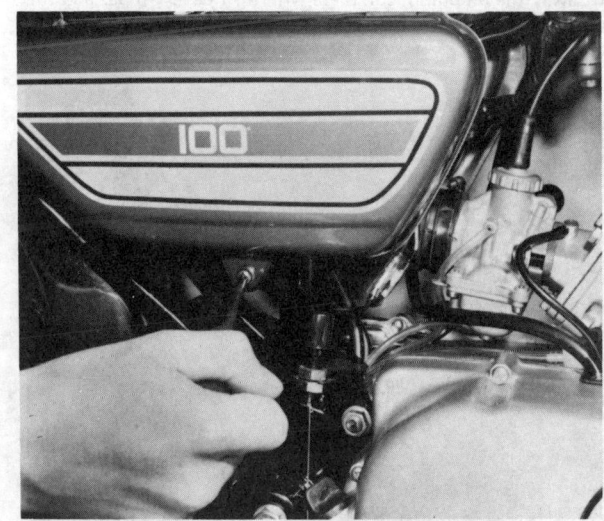

11.1 Air cleaner case is below right-hand side panel

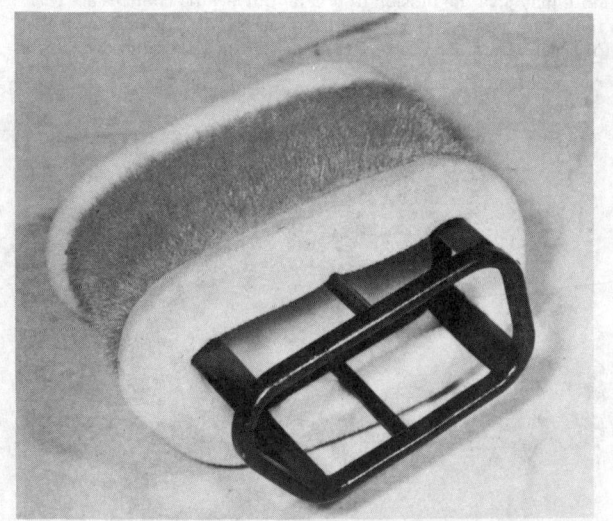

11.3 Element should be removed for washing and relubrication

## Chapter 2: Fuel system and lubrication

### 12 Crankcase drain plug

1  In common with other Yamaha models the engine of the RS100 and 125 is not fitted with a crankcase drain plug. The drain plug on the underside of the unit is used for draining the gearbox oil only.
2  Should the engine become flooded with petrol, to the extent that starting is impossible, it is best resolved by adopting the following procedure.
3  Turn the fuel tap off, and remove the spark plug. With the throttle twistgrip wide open, spin the engine over briskly with the kickstart. Replace the spark plug and lead, and attempt starting. It may be more effective to push-start the machine in second gear, as this gives a higher and more prolonged cranking speed.

### 13 Exhaust pipe and silencer: examining and cleaning

1  The exhaust system comprises a separate exhaust pipe and silencer, joined by a large gland nut. The exhaust pipe is mounted to the exhaust port via a flange and two studs. The silencer is attached rigidly to the frame by one engine mounting bolt and an outrigged bracket.
2  The parts most likely to require attention are the silencer baffles which will block up with a sludge composed of carbon and oil if not cleaned out at regular intervals. A two-stroke engine is very susceptible to this fault, which is caused by the oily nature of the exhaust gases. As this sludge builds up, back pressure will increase, with a resulting fall-off in performance.
3  There is no necessity to remove the exhaust system in order to gain access to the baffle. It is retained in the end of the silencer by a screw, reached through a hole cut in the underside of the silencer body, close to the end. When the screw is removed, the baffle can be withdrawn.
4  If the build up of carbon and oil is not too great, a wash with a petrol/paraffin mix will probably suffice as the cleaning medium. Otherwise, more drastic action will be necessary, such as the application of a blowlamp flame to burn away the accumulated deposits. Before the baffle is refitted, it must be thoroughly clean, with none of the holes obstructed.
5  When replacing the baffle, make sure the retaining screw is located correctly and tightened fully. If the screw falls out the baffle will work loose, creating excessive exhaust noise accompanied by a marked fall off in performance.
6  Do not run the machine without the baffle in the silencer or modify the baffle in any way. Although the changed exhaust note may give the illusion of greater power, the chances are that the performance will fall off, accompanied by a noticeable lack of acceleration. The carburettor is jetted to take into account the fitting of a silencer of a certain design and if this balance is disturbed the carburation will suffer accordingly.

### 14 The lubrication system

1  Unlike many two-strokes, the Yamaha RS models have an independent lubrication system for the engine and do not require the mixture of a measured quantity of oil to the petrol content of the fuel tank in order to utilise the so-called 'petroil' method. Oil of the correct viscosity (SAE 30) is contained in a separate oil tank mounted on the left-hand side of the machine and is fed to a mechanical oil pump on the right-hand side of the engine which is driven from the crankshaft by reduction gear. The pump delivers oil at a predetermined rate, via two flexible plastic tubes, to oilways in the inlet passage of the cylinder barrel. In consequence, the oil is carried into the engine by the incoming charge of petrol vapour, when the inlet port opens.
2  The oil pump is also interconnected to the twist grip throttle, so that when the throttle is opened, the oil pump setting is increased a similar amount. This technique ensures that the lubrication requirements of the engine are always directly related to the degree of throttle opening. This facility is arranged by means of a control cable looped around a pulley on the end of the pump. The cable is joined to the throttle cable junction box.

### 15 Removing and replacing the oil pump

1  There is no necessity to remove the oil pump assembly unless the cover itself is damaged and has to be replaced. Under these circumstances the oil pump must be removed as a complete unit, so that it can be fitted to the new crankcase cover.
2  To remove the right-hand crankcase cover, slacken the cross-headed screws retaining it to the engine casing, having first removed the oil pump cover and detached the cable from the operating pulley. There is no necessity to remove the engine from the frame in order to complete this operation, but the oil delivery pipes must be detached from the cylinder barrel.
3  The oil pump is secured to the cover by two cross-headed screws; but before these can be removed, the drive pinion must first be detached. This is located on the inside of the outer cover and is held by a nut and spring washer. When both are removed, the plastic pinion can be pulled off the oil pump drive shaft and the short metal rod passing through the shaft removed.
4  Unscrew the two cross-head screws from the other side of the crankcase cover, remove the circlip from the drive spindle, and lift away the oil pump, complete with the delivery pipes and the grommet through which they pass at the top of the crankcase cover.

13.3 Baffle may be withdrawn after retaining screw is removed

14.1 Oil is injected into inlet tract via this union

## Chapter 2: Fuel system and lubrication

5 Refit the oil pump to the replacement crankcase cover, using a new gasket at the oil pump/crankcase cover joint and a new oil seal behind the drive pinion. Replace and tighten the two cross-head mounting screws. The remainder of the reassembly is accomplished by reversing the dismantling procedure, but do not replace the front portion of the crankcase cover because the oil pump must be bled to ensure the oil lines are completely free from air bubbles.

### 16 Bleeding the oil pump

1 It is necessary to bleed the oil pump every time the main feed pipe from the oil tank is removed and replaced. This is because air will be trapped in the oil line, no matter what care is taken when the pipe is removed.
2 Check that the oil pipe is connected correctly, with the retaining wire clip in position. Then remove the cross-head screw with the fibre washer beneath the head from the outer face of the pump body. This is the oil bleed screw.
3 Check the oil in the tank is not close to the refill level, then place a container below the oil bleed hole to collect the oil that is expelled as the pump is bled. Rotate the white plastic pinion at the base of the oil pump in a clockwise direction; this is the pinion with a milled edge with arrows showing the direction of rotation stamped on the face. Continue rotating the pinion until the oil expelled from the bleed hole is completely free from air bubbles, then replace the bleed screw and fibre washer. DO NOT replace the front portion of the crankcase cover until the pump setting has been checked, as described in the next Section.
4 Late models may be found to be fitted with a modified oil pump with no bleed pinion. In this case, remove the bleed screw, open the throttle twistgrip fully, and spin the engine by way of the kickstart lever. Continue until the stream of oil from the bleed hole is free of air bubbles, then replace the bleed screw and washer.

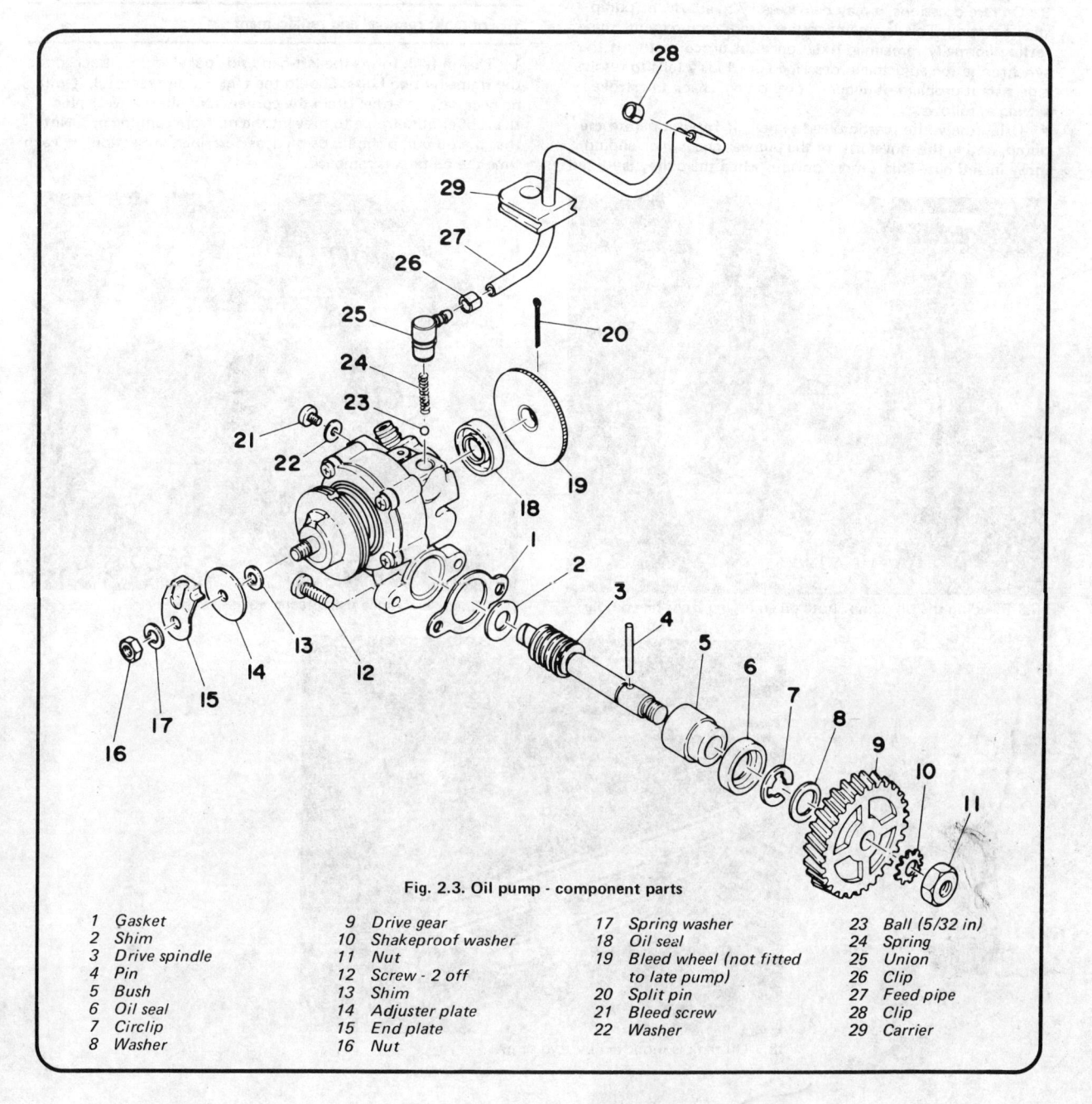

Fig. 2.3. Oil pump - component parts

| 1 | Gasket | 9 | Drive gear | 17 | Spring washer | 23 | Ball (5/32 in) |
| 2 | Shim | 10 | Shakeproof washer | 18 | Oil seal | 24 | Spring |
| 3 | Drive spindle | 11 | Nut | 19 | Bleed wheel (not fitted | 25 | Union |
| 4 | Pin | 12 | Screw - 2 off |  | to late pump) | 26 | Clip |
| 5 | Bush | 13 | Shim | 20 | Split pin | 27 | Feed pipe |
| 6 | Oil seal | 14 | Adjuster plate | 21 | Bleed screw | 28 | Clip |
| 7 | Circlip | 15 | End plate | 22 | Washer | 29 | Carrier |
| 8 | Washer | 16 | Nut |  |  |  |  |

## 17 Checking the oil pump settings

1  The area most likely to require attention is the pump control cable adjustment. This setting is important, as it governs the ratio of oil to fuel mixture fed to the engine. Before the pump control cable is adjusted, check that the lower throttle cable is adjusted to give 0.5 – 1.0 mm (0.020 – 0.039 in) free play. The throttle cable clearance is set by means of the adjuster on the carburettor top.

2  Slowly turn the throttle twistgrip to establish the point at which all free play is removed from the throttle cable. At this point, the throttle valve should be just about to commence movement. Check the position of the circular alignment mark on the face of the pump pulley. This should align with the protruding guide pin. If necessary, the pump control cable should be adjusted to obtain the correct setting.

3  On rare occasions, it may be necessary to adjust the pump stroke setting. This is a comparatively rare requirement, the setting normally remaining fixed once set up correctly. If the synchronisation adjustment described previously fails to resolve a persistant problem of under or over oiling, check the stroke setting as follows.

4  Using the white plastic bleed pinion, if fitted, operate the pump, noting the movement of the plunger end, which should move in and out. This is more obvious when the pulley is turned by opening the throttle twistgrip. On later pumps, no bleed pinion is fitted, and in this case the engine must be turned over with the kickstart to operate the pump. This is a laborious process, and it can be eased somewhat by removing the sparking plug. Arrange the plunger so that it is fully outward, indicating that the pump is at minimum stroke. Close the throttle, and measure the gap between the adjuster plate and casing using feeler gauges. Note the reading, then repeat the operation once or twice to ensure that a maximum gap reading is obtained.

5  The gap reading should be within the range 0.20 – 0.25 mm (0.008 – 0.010 in) if the pump stroke setting is correct. If this is not the case, remove the nut, spring washer and adjuster plate, and remove or fit shims to obtain the correct clearance. 0.1 mm (0.004 in) shims are available from Yamaha Service Agents. Always re-check the setting after refitting the adjuster plate and tightening the securing nut.

## 18 Oil tank: removal and replacement

1  The oil tank forms the left-hand side panel and is attached to the frame by two bolts. Should the tank require removal, it will be necessary to either drain the contents, or alternatively plug the rubber outlet pipe to prevent the oil from running out. Note that the oil pump should be bled, as described in Section 16, each time the oil tank is removed.

16.3 Bleeding the oil pump: note oil emerging from bleed hole

17.2 Check clearance using feeler gauge

18.1 Oil tank is mounted by two screws

## Chapter 2: Fuel system and lubrication

## 19 Fault diagnosis: fuel system

| Symptom | Cause | Remedy |
| --- | --- | --- |
| Excessive fuel consumption | Air cleaner choked or restricted | Clean or renew element. |
| | Fuel leaking from carburettor | Check all unions and gaskets. |
| | Badly worn or distorted carburettor | Renew. |
| | Carburettor settings incorrect | Readjust. Check settings with specifications. |
| Idling speed too high | Throttle stop screw in too far | Adjust screw. |
| | Carburettor top loose | Tighten. |
| Engine sluggish. Does not respond to throttle | Back pressure in silencer | Check baffle and clean if necessary. |
| Engine dies after running for a short while | Blocked vent hole in filler cap | Clean. |
| | Dirt or water in carburettor | Remove and clean. |
| General lack of performance | Weak mixture; float needle sticking in seat | Remove float chamber and check needle seating. |
| | Air leak at carburettor or leaking crankcase seals | Check for air leaks or worn seals. |

## 20 Fault diagnosis: lubrication system

| Symptom | Cause | Remedy |
| --- | --- | --- |
| White smoke from exhaust | Too much oil | Check oil pump setting and reduce if necessary. |
| Engine runs hot and gets sluggish when warm | Too little oil | Check oil pump setting and increase if necessary. |
| Engine runs unevenly, not particularly responsive to throttle openings | Intermittent oil supply | Bleed oil pump to displace air in feed pipe. |
| Engine dries up and seizes | Complete lubrication failure | Check for blockages in feed pipe, also whether oil pump drive has sheared |

# Chapter 3 Ignition system

*For information relating to 1977 on models refer to Chapter 7*

## Contents

| | |
|---|---|
| General description ... 1 | Condenser: testing ... 6 |
| Ignition system: methods of checking ... 2 | Ignition coil: checking ... 7 |
| Contact breaker: gap setting and timing check ... 3 | Ignition switch ... 8 |
| Contact breaker: removal, renovation and replacement ... 4 | Spark plug: checking and resetting the gap ... 9 |
| Condenser: location, removal and replacement ... 5 | Fault diagnosis: ignition system ... 10 |

## Specifications

**Generator** | **RS100** | **RS125**
--- | --- | ---
Make ... | Mitsubishi | Hitachi
Model ... | FOOOTO2771 | F130 - 09
Output ... | 6 volts | 6 volts

**Ignition coil**
Type ... External - frame mounted

**Spark plug**
Make ... NGK
Type ... B - 8HS
Reach ... 13 mm (½ in)
Gap ... 0.5 - 0.6 mm (0.020 - 0.024 in)

## 1 General description

1 The spark necessary to ignite the petrol/air mixture in the combustion chamber is derived from a flywheel generator mounted on the left-hand side of the crankshaft. The ignition coil is attached to the frame beneath the fuel tank, low tension current being fed to its primary windings from the battery, when the ignition switch is on.

2 The contact breaker assembly forms a switch mechanism actuated by a lobe on the contact breaker cam. As the points separate, low tension current in the primary windings of the ignition coil is interrupted, and high tension current is induced in the secondary windings, causing a spark to jump across the spark plug electrodes.

3 As mentioned previously, charging coils are included in the generator to supply the electrical needs of the machine. The AC current produced passes through a silicon rectifier which converts it to DC, and from there it passes to the battery and electrical system. The two coils, along with the contact breaker assembly, are mounted on an aluminium alloy stator plate which is attached to the outside of the left-hand crankcase.

## 2 Ignition system: methods of checking

1 It is quite possible to check the ignition system using a small multimeter test instrument, and reference is made to these tests throughout this Chapter. A multimeter is comparatively inexpensive, and well worth obtaining, due to its versatility. It is assumed that anyone possessing one will have some experience in its use. Where possible, checking procedures are given which avoid the use of any type of meter, but it must be stressed that these checks can not give a diagnosis with the degree of accuracy possible with a test meter. It follows that, if test facilities are not available, any rough checks should be backed up by taking the component concerned to a Yamaha Service Agent for verification rather than instantly condemning it.

## 3 Contact breaker: gap setting and timing check

### A Contact breaker gap

1 It will be noticed from the Section heading that contact breaker adjustment has been grouped together with ignition timing checking. This is because the only form of timing adjustment possible with this type of ignition generator is that of varying the contact breaker points gap within certain limits.

2 With the left-hand outer casing removed, it is possible to view the contact breaker assembly through the windows in the flywheel. Remove the spark plug, and turn the flywheel until the points are seen to separate. Continue turning until the contact breaker is fully open. At this stage the contact breaker points faces should be examined to see if dressing or renewal is required. Should this be the case, the necessary remedial action must be carried out before any adjustment is made, following the procedure given in Section 4 of this Chapter.

3 Assuming the contact breaker to be in good condition, check the gap when fully open, using metric feeler gauges if possible. The correct setting is 0.3 - 0.4 mm (0.012 - 0.016 in). If necessary, adjust the gap by moving the fixed contact, after slackening the retaining screw just enough to permit movement

# Chapter 3: Ignition system

of the base plate. Tighten the retaining screw and recheck the setting.

*B Checking the ignition timing*

4 With the contact breaker gap set correctly, as described above, the timing can be checked using a DTI (dial test indicator) fitted into a 14 mm spark plug adaptor. Mount the DTI in the spark plug hole, and rotate the engine to find TDC (top dead centre), when the dial will indicate its highest reading. Zero the DTI scale at this point, and then turn the engine in its normal direction of rotation (clockwise). Obtain one of the following items of test equipment to determine contact breaker point separation: a self-powered points checker, a battery and bulb test circuit (Fig. 3.1) or a multimeter set to the x 1 ohm scale. Connect the negative (−) probe of the test equipment to a good earth point on the engine and the positive (+) probe to the black output lead from the generator. The exact point of separation will be shown by the gauge needle swinging from closed to open on the points checker, by the bulb dimming if using the battery and bulb method, or by flickering of the multimeter needle, indicating increased resistance.

5 Turn the engine gradually until the points just separate (indicated by the meter or bulb) and note the reading on the DTI scale. The ignition timing is correct if separation occurs within 1.3 - 2 mm (0.0512 - 0.0787 in) before top dead center (BTDC). It is possible to adjust the contact breaker gap to achieve separation at this point, but it should be noted that both the contact breaker gap and the ignition setting should be kept within the limits given for each. Should this not be possible, it will be due to wear, either of the contact faces themselves or of the fibre heel of the moving contact. In instances such as these, the contact breaker assembly must be renewed.

## 4 Contact breaker: removal, renovation and replacement

1 If on examination, the contact faces appear burnt or pitted, they must be removed for refacing or renewal, depending on the extent of the damage. The manufacturers recommend that a flywheel holding tool is used for slackening the securing nut. If this is not to hand, select top gear and apply the rear brake to immobilise the crankshaft whilst the nut is removed.

2 The flywheel is keyed to a taper, and in order to break the taper joint it will be necessary to use either the manufacturer's threaded puller, or, if this is not available, an ordinary two-legged universal puller. If using the latter, place a spacer, in the form of an unwanted nut, on the mainshaft end to prevent damage to the thread. Tighten the puller gradually, tapping the end to free the flywheel from the mainshaft, then remove the flywheel and its Woodruff key and place them to one side.

3 The contact breaker assembly is retained by a single screw, and is easily removed after disconnecting the cable from its terminal. Using a small electrical screwdriver, prise off the circlip which retains the moving contact assembly to its pivot pin. Remove the plain washer, followed by the moving contact complete with insulating washers. Make a note of the order in which components are removed, as they are easily assembled incorrectly.

4 The points surfaces may be dressed by rubbing them on an oilstone or fine emery paper, keeping the points square to the abrasive surface. If possible, finish off by using Crokus paper to give a polished surface, which is less prone to subsequent pitting. Make sure all traces of abrasive are removed before reassembly. If however, it is necessary to remove a substantial amount of material before the faces can be restored, new contacts should be fitted. If this precaution is not observed, it may prove impossible to obtain the correct ignition timing and contact breaker gap (See the preceding Section for details).

5 Reassemble the contact breaker assembly by reversing the dismantling sequence, taking care that the insulating washers are replaced correctly. If this precaution is not observed, it is easy to inadvertently earth the assembly rendering it inoperative. The pivot pin should be greased sparingly, and a few drops of oil applied to the cam lubricating wick.

6 If the contact breaker is being renewed due to excessive burning of the contacts, this is likely to have been caused by a faulty condenser. Refer to the next Section if this is suspected.

3.2 Remove outer cover to gain access to generator

3.3a Slacken retaining screw slightly ...

3.3b ... to enable contact to be moved using screwdriver

## Chapter 3: Ignition system

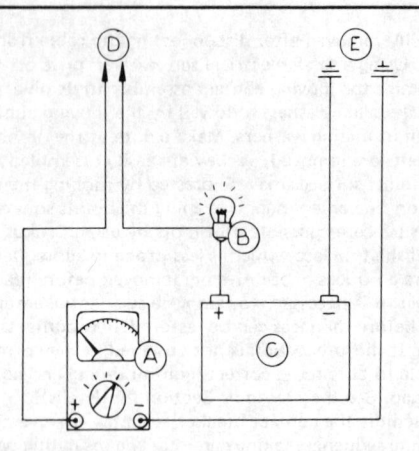

**Fig. 3.1. Using multimeter or battery and bulb arrangement to establish contact breaker separation**

A  Multimeter set on resistance mode
B  High-wattage bulb
C  Battery
D  Probe to moving contact terminal
E  Earth

### 5  Condenser: location, removal and replacement

1  A condenser is included in the contact breaker circuit to prevent arcing across the contact breaker points as they separate. It is connected in parallel with the contact points, and if a fault develops in the condenser, ignition failure is liable to occur.
2  If the engine is difficult to start, or if misfiring occurs, it is possible that the condenser is at fault. To check whether it has failed, observe the points whilst the engine is running. If excessive sparking occurs across the points faces, and they have a blackened or burnt appearance, it may be assumed that the condenser is no longer serviceable.
3  The condenser is located on the bottom of the stator, and is retained by a single screw through the strap soldered to the body of the condenser and by the lead wire attached to the screw and nut passing through the end of the moving contact return spring. Remove the screw and nut so that the terminal end is freed. Because it is impracticable to repair a defective condenser, a new one must be fitted.

### 6  Condenser: testing

1  Without the appropriate test equipment, there is no alternative means of verifying whether a condenser is still serviceable.
2  Bearing in mind the low cost of a condenser, it is far more satisfactory to check whether it is malfunctioning by direct replacement.

### 7  Ignition coil: checking

1  If ignition problems are present, and the contact breaker and condenser have been eliminated, attention should be turned to the coil. It is bolted to the frame beneath the fuel tank, which may be removed to gain better access, if required.
2  A multimeter is again the best instrument to use for checking the coil windings. Set the meter on the resistance setting and connect the negative lead to the coil frame (earth). Remove the spark plug cap and disconnect the low tension cable at the bullet connector. Place the positive probe against the low tension lead first and note the reading (Primary coil resistance). Repeat the test, placing the positive probe against the high tension spark plug lead (Secondary coil resistance).
3  The resistance readings should be as follows:

|  | RS100 | RS125 |
|---|---|---|
| Primary coil resistance | 1.02 ohms $\pm$ 10% | 1.7 ohms $\pm$ 10% |
| Secondary coil resistance | 6.0k ohms $\pm$ 10% | 6.0k ohms $\pm$ 10% |

Readings should be taken at 20°C (68°F).

Slight variation may be encountered if the ambient temperature departs greatly from that given. If the values differ from those given, the coil is faulty.
4  If a multimeter is not available, and by means of testing, the other components have been found to be satisfactory, the following method may be used to give an estimation of the coil's condition. Remove the suppressor cap and bare the inner wire. Remove the contact breaker cover and turn the engine over until the contact breaker points relevant to the coil to be tested are closed. Turn the ignition on and using an insulated screwdriver flick the points open and shut. If the bared end of the HT lead is held approximately 5 mm from an earthing point (the cylinder head) whilst this is done, a blue spark should jump the gap. If the spark is unable to jump a gap, or is yellowish in colour, the coil is probably at fault.
5  The ignition coil is a sealed unit and it is not possible to effect a satisfactory repair in the event of failure. A new coil must be fitted.

### 8  Ignition switch

1  The ignition switch is a multi-point switch which also controls the lighting circuits. It is bolted from the underside to the top yoke of the forks and is located immediately in front of the handlebars.
2  The switch is unlikely to malfunction during the normal service life of the machine and does not require any maintenance. If, however, an ignition fault appears to lie in the switch, it should be removed and tested. This is covered in Chapter 6 of this Manual.

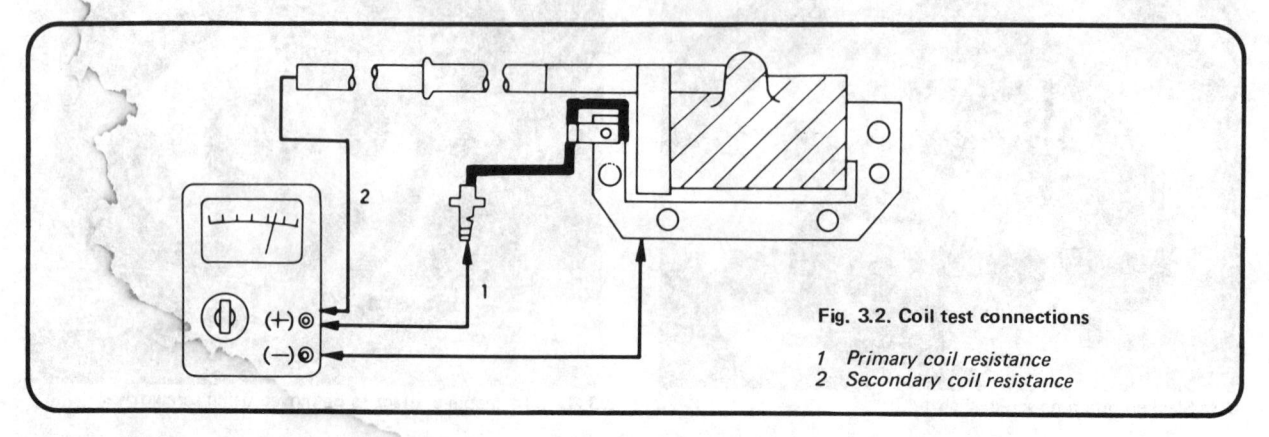

**Fig. 3.2. Coil test connections**

1  Primary coil resistance
2  Secondary coil resistance

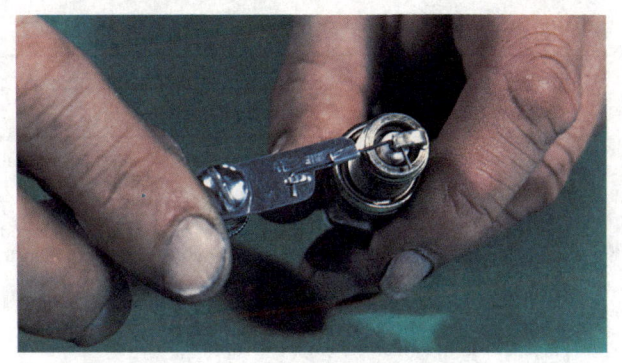

**Spark plug maintenance:** Checking plug gap with feeler gauges

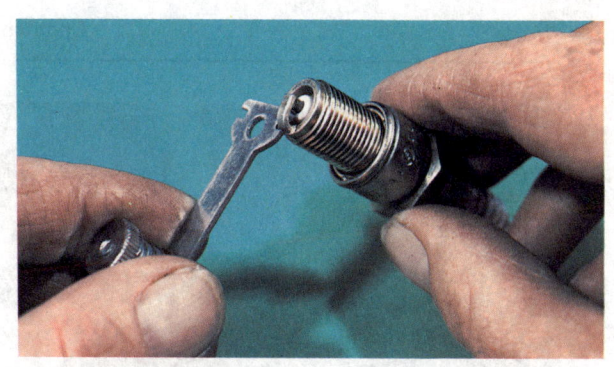

Altering the plug gap. Note use of correct tool

**Spark plug conditions:** A brown, tan or grey firing end is indicative of correct engine running conditions and the selection of the appropriate heat rating plug

White deposits have accumulated from excessive amounts of oil in the combustion chamber or through the use of low quality oil. Remove deposits or a hot spot may form

Black sooty deposits indicate an over-rich fuel/air mixture, or a malfunctioning ignition system. If no improvement is obtained, try one grade hotter plug

Wet, oily carbon deposits form an electrical leakage path along the insulator nose, resulting in a misfire. The cause may be a badly worn engine or a malfunctioning ignition system

A blistered white insulator or melted electrode indicates over-advanced ignition timing or a malfunctioning cooling system. If correction does not prove effective, try a colder grade plug

A worn spark plug not only wastes fuel but also overloads the whole ignition system because the increased gap requires higher voltage to initiate the spark. This condition can also affect air pollution

## Chapter 3: Ignition system

7.1 Coil is partially hidden below top tube

### 9 Spark plug: checking and re-setting the gap

1  An NGK B-8HS 14 mm spark plug is fitted to all RS models as standard equipment.
2  The recommended gap is 0.5 - 0.6 mm (0.020 - 0.024 in). The gap can be reset by carefully bending the outer, earth, electrode to obtain the required setting. The clearance is correct when the appropriate feeler gauge is a light sliding fit between the electrodes. Never bend the central electrode, otherwise the insulator will crack, causing engine damage if the broken particles fall in whilst the engine is running.
3  After some experience the spark plug electrodes can be used as a reliable guide to engine operating conditions. See photographs on page 59.
4  Always carry a spare spark plug of the correct type. The plug in a two-stroke engine lead a particularly hard life and are liable to fail more readily than when fitted to a four-stroke.
5  Never overtighten a spark plug, otherwise there is risk of stripping the threads from the cylinder head, especially as it is cast in light alloy. A stripped thread can be repaired without having to scrap the cylinder head by using a 'Helicoil' thread insert. This is a low-cast service, operated by a number of dealers.
6  Before replacing a spark plug into the cylinder head, coat the threads sparingly with a graphited grease to aid future removal. Use the correct sized spanner when tightening a plug, otherwise the spanner may slip and damage the ceramic insulator. The plug should be tightened sufficiently to seat firmly on its sealing washer, and no more.
7  Make sure that the plug insulating cap is a good fit and free from cracks. Apart from acting as an insulator from water and road dirt it contains the suppressor for eliminating radio and TV interference.

### 10 Fault diagnosis: ignition system

| Symptom | Cause | Remedy |
| --- | --- | --- |
| Engine will not start | No spark at plug | Faulty ignition switch. Check whether current is reaching ignition coil. |
|  | Weak spark at plug | Dirty contact breaker points require cleaning. Contact breaker gap has closed up. Re-set. |
| Engine starts, but runs erratically | Intermittent or weak spark | Renew spark plug. If no improvement check whether points are arcing. If so renew condenser. |
|  | Ignition over-advanced | Check ignition timing and if necessary, re-set. |
|  | Plug lead insulation breaking down | Check for breaks in outer covering, especially near frame. |
| Engine difficult to start and runs sluggishly. Overheats | Ignition timing retarded | Check ignition timing/contact breaker setting. |

# Chapter 4 Frame and forks

*For information relating to 1977 on models refer to Chapter 7*

**Contents**

| | |
|---|---|
| General description ... 1 | Centre stand: examination ... 14 |
| Front forks: removal - general ... 2 | Prop stand: examination ... 15 |
| Front forks: removal from frame ... 3 | Footrests: examination ... 16 |
| Steering head bearings: examination and renovation ... 4 | Rear brake pedal: examination and renovation ... 17 |
| Front fork legs: removal from yokes ... 5 | Dualseat: removal and replacement ... 18 |
| Front fork legs: dismantling (disc brake models) ... 6 | Kickstart lever: examination and renovation ... 19 |
| Front fork legs: dismantling (drum brake models) ... 7 | Speedometer and tachometer heads: removal and replacement ... 20 |
| Front forks: examination and renovation ... 8 | Speedometer and tachometer drive cables: examination and maintenance ... 21 |
| Front forks: reassembly and replacement ... 9 | |
| Steering head lock ... 10 | Speedometer and tachometer drive: location and examination ... 22 |
| Frame: examination and renovation ... 11 | |
| Swinging arm rear fork: removal, examination and renovation ... 12 | Cleaning the machine ... 23 |
| Rear suspension units: examination ... 13 | Fault diagnosis: frame and forks ... 24 |

**Specifications**

### Front forks
| | |
|---|---|
| Type ... | Oil damped telescopic |
| Damping oil capacity ... | 150 cc (all models) |
| Damping oil specification ... | Proprietary fork oil or SAE 30 engine oil |

### Rear suspension
| | |
|---|---|
| Type ... | Swinging arm |

### Suspension units
| | |
|---|---|
| Type ... | Oil damped coil spring. Sealed oil content |

## 1 General description

The Yamaha RS models employ a conventional tubular cradle frame. The front forks are telescopic, two types being fitted, dependant on whether the machine has a drum or disc front brake arrangement. Rear suspension is provided by a swinging arm fork, pivoting on replaceable bushes, and controlled by hydraulically damped rear suspension units.

## 2 Front forks: removal: general

1 It is unlikely that the forks will require removal from the frame unless the fork seals are leaking or accident damage has been sustained. In the event that the latter has occurred, it should be noted that the frame may also have become bent, and whilst this may not be obvious when checked visually, could prove to be potentially dangerous.
2 If attention to the fork legs only is required, it is unnecessary to detach the complete assembly, the legs being easily removed individually.
3 If, on the other hand, the headstock bearings are in need of attention, the forks complete with bottom yoke must be removed.
4 Before any dismantling work can be undertaken, the machine should be placed on the centre stand, and blocked securely so that the front wheel is held off the ground. Detach the speedometer drive cable at the wheel, by unscrewing the knurled gland nut which retains it. On drum brake wheels, slacken the wheel spindle pinch bolt, disc brake models have a corresponding spindle clamp, which should be released by slackening the two bolts which retain it. Remove the front brake cable from drum brake models.
5 Remove the wheel spindle nut and split pin, and withdraw the spindle with the aid of a tommy bar. The wheel can now be lowered clear of the forks and put to one side.
  To avoid damage to the paintwork, it is a good idea to remove the front mudguard at this stage, irrespective of whether removal will later be necessary, as in the case of individual fork leg removal. The brake caliper, on disc brake models, shares two of the four mudguard mounting bolts. This too can be pulled clear of the forks.
  Note that if the fork legs are to be dismantled it is advisable, though not necessary, to drain the damping oil at this stage. The legs can be pumped up and down to force the oil out of the drain hole. Each leg contains 150 cc of oil.

## 3 Front forks: removal from frame

1 Having detached the front wheel, as described in the preceding Section, the handlebars and controls must be removed.
2 Disconnect one of the battery terminals, to make sure that no shorting occurs when the switch and headlamp units are removed.
3 Remove the clutch lever complete with cable, and also the left-hand switch unit. Slacken the two screws clamping the throttle twistgrip and switch unit, and slide it off the handlebar end. Remove the clamp bolts from the master cylinder assembly on disc brake models and pull it clear. Do not allow it to hang from its hose; it should be tied up to the frame.
4 Slacken and remove the four handlebar clamp bolts and remove the clamps complete with the handlebars. Remove the two bolts retaining the headlamp shell. Unscrew the knurled rings retaining the speedometer and tachometer cables, and free the cables. Detach the instrument heads from their brackets.
5 Remove the chromed dome nut from the top fork yoke, and unscrew and remove the chrome fork nuts. The top yoke can now be pulled away, complete with the instrument brackets. Remove the guides which route the hydraulic hose (disc brake models) and the speedometer cable.
6 Using a C spanner, slacken the slotted nut on the top of the headstock assembly. As the assembly is lowered, the uncaged balls of the lower race will drop free. These should be caught, there being 19 balls in the lower race.
7 Draw the fork, bottom yoke and steering column downwards to disengage it from the steering head, threading the hydraulic hose around the stanchion (disc brake models). It may be necessary to raise the front of the machine to give sufficient clearance for the fork assembly to be pulled clear.
8 Removal of the drum brake type fork assembly is essentially the same as that outlined in the foregoing Section, the obvious exception being the hydraulic system.
9 The front brake lever should be removed from the handlebars in this case. The cable, having been detached when the front wheel was removed, can be withdrawn with the lever.

## 4 Steering head bearings: examination and renovation

1 Before reassembly of the forks is commenced, examine the steering head races. The ball bearing tracks of the respective cup and cone bearings should be polished and free from indentations or cracks. If wear or damage is evident, the cups and cones must be renewed as a complete set. They are a tight press fit and should be drifted out of position.
2 Ball bearings are cheap. If the originals are marked or discoloured, they should be renewed. To hold the steel balls in position during reassembly, pack the bearings with grease. Note that each race contains only 19 or 22 ball bearings. There is space for the addition of one extra ball, but this must be left empty to prevent the ball bearings from skidding on one another, a situation which would greatly accelerate the rate of wear.

## 5 Front fork legs: removal from yokes

1 The two types of fork fitted to the RS models are quite different in construction, although they can be dealt with in a similar manner for the purposes of removal. If the front fork legs only require attention, they can, as mentioned earlier, be removed individually. This avoids extra dismantling work.
2 With the machine supported, and the front wheel removed, as detailed in Section 2, paragraph 4 and 5 of this Chapter, proceed as follows:
3 Deal with one leg at a time, as this obviates the risk of transferring components from one side to the other. On disc brake models, the caliper unit should be pulled clear as the

5.4 Slacken bolts to release mudguard assembly

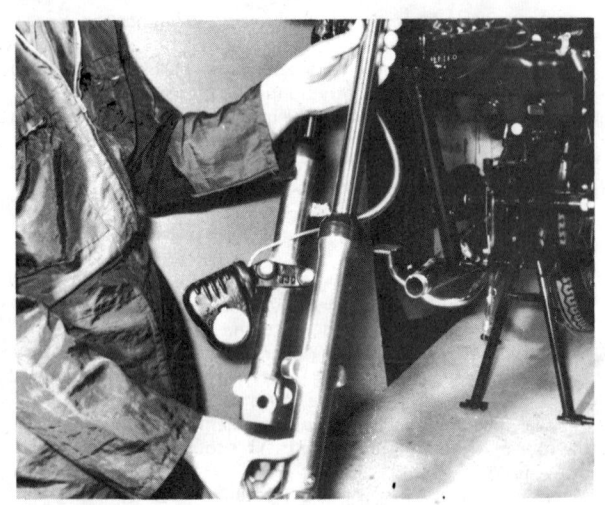

5.5a Either type of fork can be pulled clear ...

5.5b ...after slackening clamp bolts and top bolts

## Chapter 4: Frame and forks

mudguard is removed.

4 Remove the four bolts securing the front mudguard, and on disc brake models, the caliper unit. Unscrew the chromium plated fork top nuts, and slacken the pinch bolt on the bottom yoke. On later models without fork shrouds, slacken the pinch bolt which clamps the headlamp bracket to the stanchion.

5 Withdraw the fork leg by pulling downwards. If necessary, the fork top nut may be partially refitted, and then tapped lightly with a hide mallet to jar the leg loose.

### 6 Front fork legs: dismantling (disc brake models)

1 If not carried out previously, remove the drain plug at the bottom of each leg and allow the oil to drain into a suitable container, each leg holding 150 cc of oil. This operation is facilitated by pumping the fork up and down, forcing oil from the damper. Replace the drain plug.

2 Remove the spring seat and spring. This is preceded by a spacer.

3 Clamp the lower fork leg in a vice, having first packed the jaws with rag to avoid damaging the surface finish. Do not overtighten the vice.

4 Pull the dust seal free of the lower leg and slide it off the stanchion. On certain types of fork the stanchion and damper assembly is retained by a bolt which passes up through the base of the lower leg. It may prove necessary to temporarily replace the spring and top bolt in order to slacken the bolt. When this has been removed, the stanchion and damper assembly can be withdrawn, and the damper assembly shaken out of the top of the stanchion.

5 On the other type of fork, remove the circlip from the top of the lower leg. The stanchion can be jarred free from the lower leg by pulling the two apart smartly, thus displacing the upper bush and oil seal. Should this method prove unsuccessful, it will be necessary to construct a form of slide hammer, using a long bolt which will thread into the top of the stanchion, and a suitable weight. No further dismantling is possible.

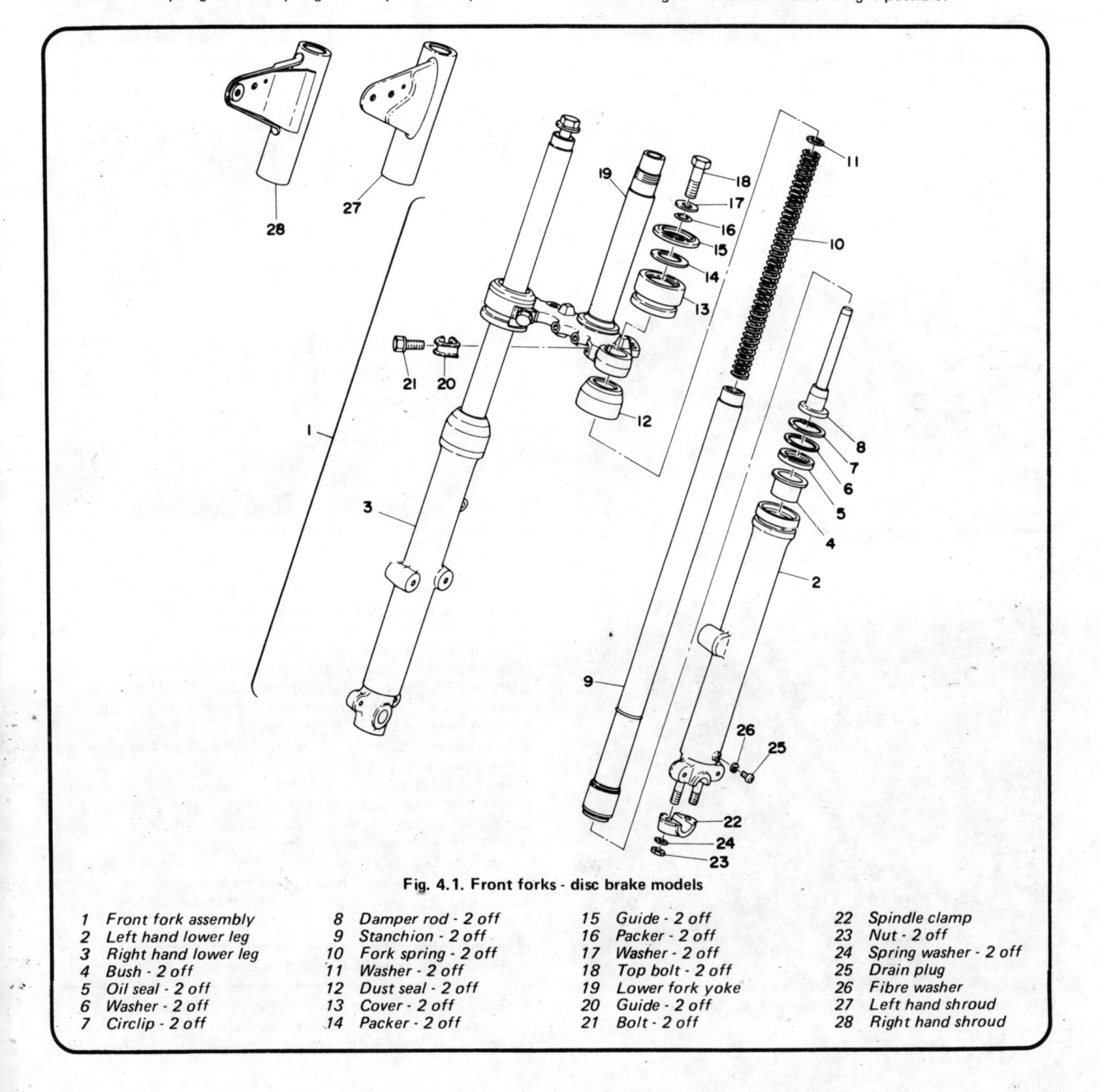

Fig. 4.1. Front forks - disc brake models

| | | | |
|---|---|---|---|
| 1 Front fork assembly | 8 Damper rod - 2 off | 15 Guide - 2 off | 22 Spindle clamp |
| 2 Left hand lower leg | 9 Stanchion - 2 off | 16 Packer - 2 off | 23 Nut - 2 off |
| 3 Right hand lower leg | 10 Fork spring - 2 off | 17 Washer - 2 off | 24 Spring washer - 2 off |
| 4 Bush - 2 off | 11 Washer - 2 off | 18 Top bolt - 2 off | 25 Drain plug |
| 5 Oil seal - 2 off | 12 Dust seal - 2 off | 19 Lower fork yoke | 26 Fibre washer |
| 6 Washer - 2 off | 13 Cover - 2 off | 20 Guide - 2 off | 27 Left hand shroud |
| 7 Circlip - 2 off | 14 Packer - 2 off | 21 Bolt - 2 off | 28 Right hand shroud |

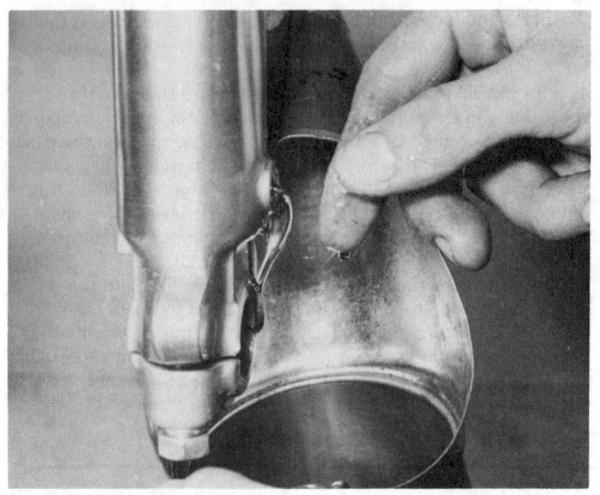

6.1 Drain off fork oil before dismantling

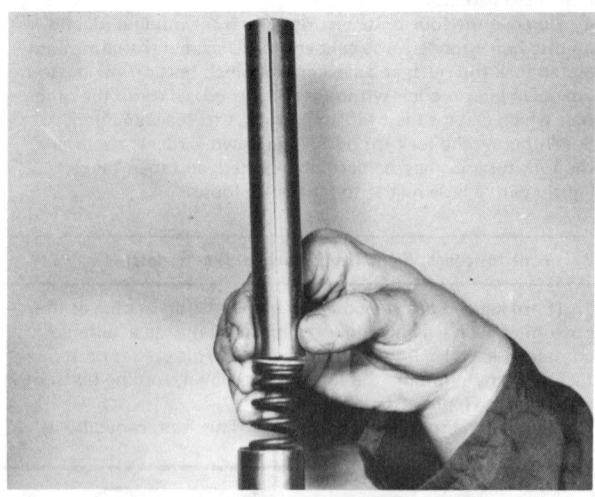

6.2 Disc brake models have internal spring and sleeve

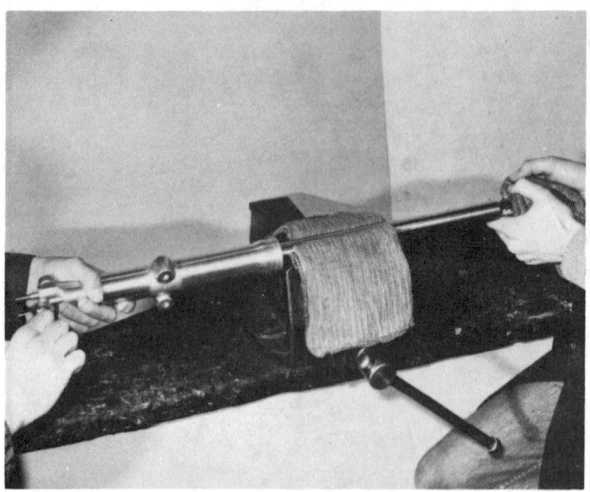

6.4a Compress spring to allow screw to be removed

6.4b Stanchion can then be pulled out

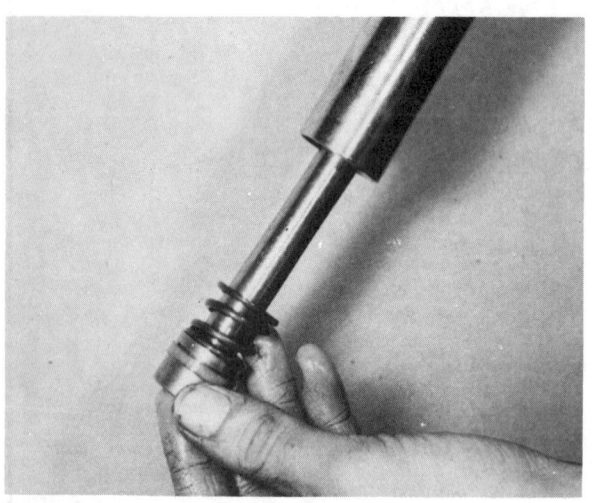

6.4c Damper components can be shaken out of stanchion

## 7 Front fork legs: dismantling (drum brake models)

1  With the fork leg(s) concerned removed, the spring shroud or gaiter and external fork spring can be slid off the stanchion together with the spring seat. The stanchion is retained in the lower leg by a long aluminium alloy sleeve nut which also supports the oil seal.

2  The sleeve nut can be removed with the aid of a strap wrench, if this is available. A good alternative method is to wrap a piece of rag around the nut to protect the polished surface, and then tighten one or two worm drive hose clips around it. A large adjustable wrench can then be used to slacken the nut, bearing against the raised boss of the clips.

3  With the nut slackened, the stanchion can be withdrawn complete with the large, headed fork bush. The various components can then be slid upwards off the stanchion.

## 8 Front forks: examination and renovation

1  The parts most liable to wear over an extended period of service are the bushes and the oil seals, especially where gaiters are not fitted. Wear is normally accompanied by a tendency for

## Chapter 4: Frame and forks

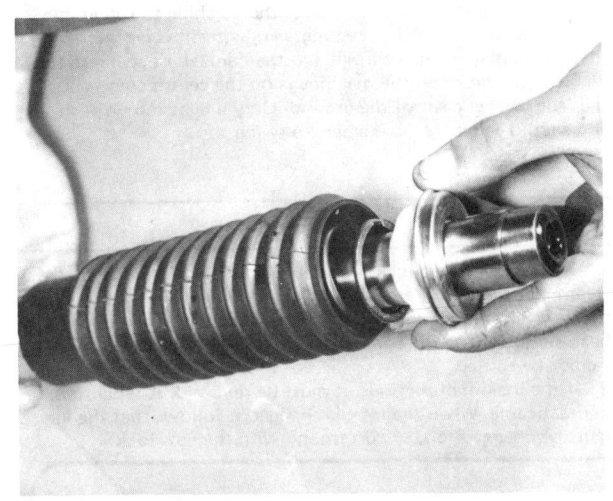

7.1a Drum brake models have external spring and gaiter

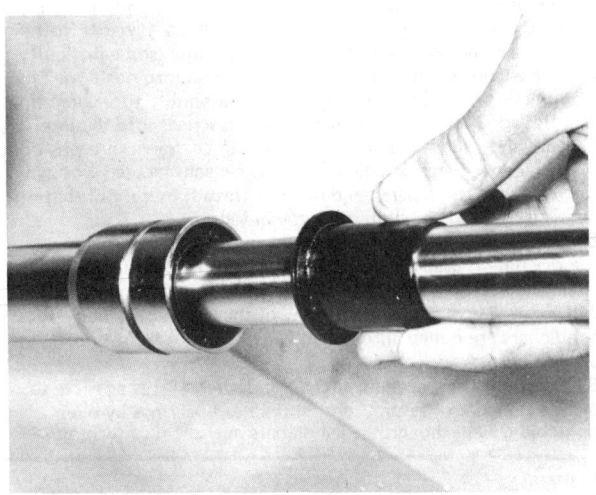

7.1b Spring seats on the nylon seat

7.2 Hose clip can be used to slacken nut

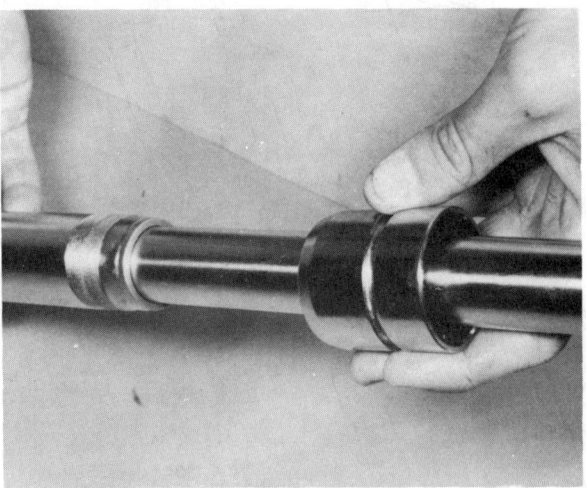

7.3 Remove nut and slide stanchion out of lower leg

the forks to judder when the front brake is applied and it should be possible to detect the increased amount of play by pulling and pushing on the handlebars when the front brake is applied fully. This type of wear should not be confused with slack steering head bearings, which can give identical results.

2  Renewal of the worn parts is quite straightforward. Particular care is necessary when renewing the oil seal. Both the seal and the fork tube should be greased to lessen the risk of damage.

3  After an extended period of service the fork springs may take a permanent set. If the overall length has decreased it is wise to fit new components. Always fit new springs as a matched pair, never separately.

4  Check the outer surface of the fork stanchion for scratches or roughness. It is only too easy to damage the oil seal during reassembly, if these high spots are not eased down. The fork stanchions are unlikely to bend unless the machine is damaged in an accident. Any significant bend will be detected by eye, but if there is any doubt about straightness, roll the stanchions on a flat surface. If the stanchions are bent, they must be renewed. Unless specialised repair equipment is available, it is rarely practicable to straighten them to the necessary standard.

5  The dust seals must be in good order if they are to fulfill their proper function. Replace any that are split or damaged.

6  Damping is effected by the damper units contained within each fork stanchion. The damping action can be controlled within certain limits by changing the viscosity of the oil used as the damping medium, although a change is unlikely to prove necessary except in extremes of climate.

### 9  Front forks: reassembly and replacement

1  Replace the front forks by reversing the dismantling sequence described in Sections 5, 6 and 7 of this Chapter. Note that it will be necessary, on DX type forks, to compress the fork assembly to hold the damper rod as the socket screw is tightened. Make sure that the backplate, or speedometer drive on DX models, locates in the slot on the fork leg as the wheel is refitted.

2  With the components in position, but not tightened down, bounce the forks up and down a few times. This ensures that each component part of the assembled unit settles into its natural position, and suffers no subsequent strain when the assembly is tightened down.

3  Tighten the unit from the wheel spindle upwards, do not omit the split pin from the wheel spindle nut. Add the recommended quantity (150 cc) of fork oil. As an alternative, an SAE

Chapter 4: Frame and forks

30 engine oil may be used. Fork oil is to be preferred, however, as it has additives in it to suppress any tendency towards frothing.

4  On machines with fork shrouds in particular, some difficulty may be encountered in drawing the stanchion into position. The manufacturers recommend the use of an assembly tool consisting of a threaded rod with a 'T' handle which screws into the internal thread on the stanchion, enabling it to be drawn into position. A more convenient expedient is to use a clean, tapered wooden rod, screwing the tapered end into the thread. Be careful that no debris is allowed to fall into the stanchion.

5  Check the adjustment of the steering head bearings before the machine is used on the road and again shortly afterwards, when they settle down. If the bearings are too slack, fork judder will occur. There should be no play at the headraces when the handlebars are pulled and pushed hard, with the front brake applied hard.

6  Overtight headraces are equally undesirable. It is possible to place a pressure of several tons on the head bearings by overtightening, even though the handlebars may seem to turn quite freely. Overtight bearings will cause the machine to roll at low speeds and give imprecise steering. Adjustment is correct if there is no play in the bearings and the handlebars swing to full lock either side when the machine is on the centre stand with the front wheel clear of the ground. Only a light tap on each end should cause the handlebars to swing.

### 10 Steering head lock

1  The steering head lock is attached to the left-hand side of the steering head. It is retained by a rivet. When in a locked position, the plunger extends and engages with a portion of the steering head stem, so that the handlebars are locked in position and cannot be turned.

2  If the lock malfunctions, it must be renewed. A repair is impracticable. When the lock is changed it follows that the key must be changed too, to correspond with the new lock.

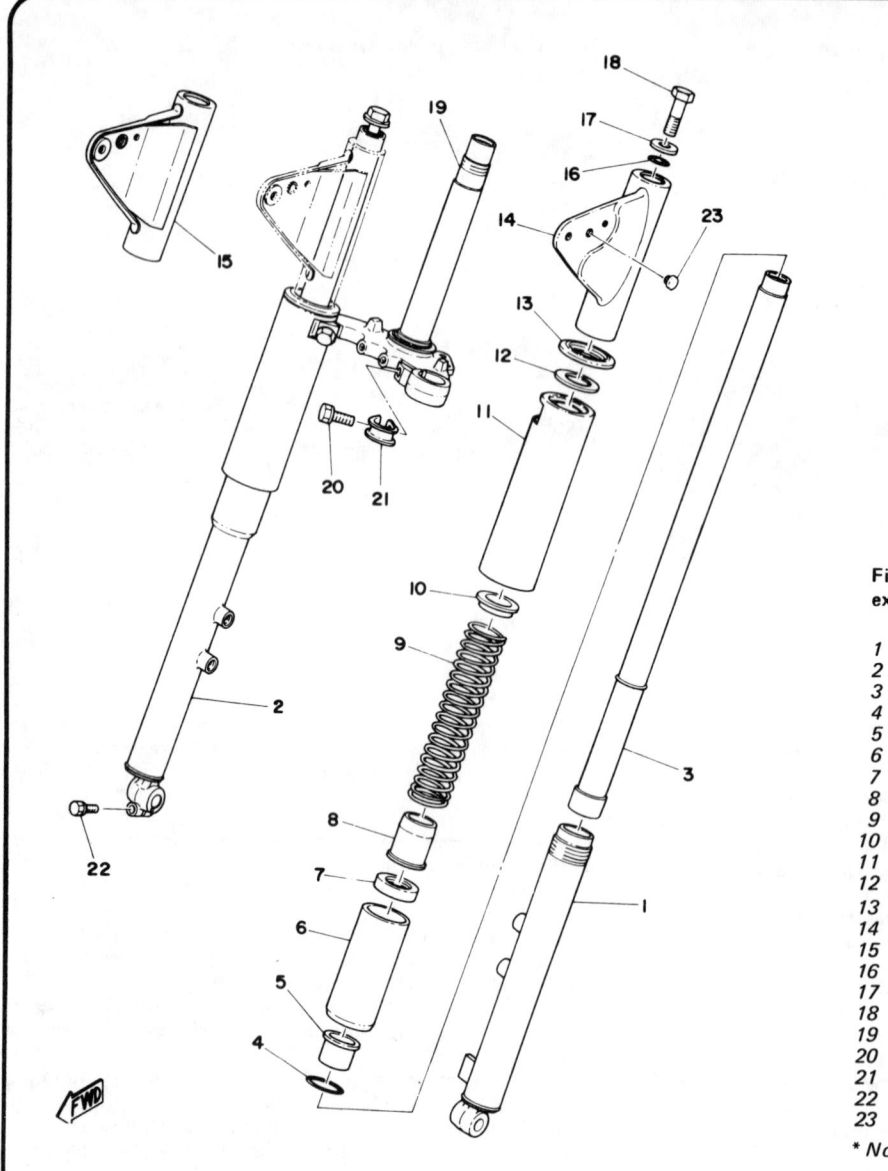

Fig. 4.2. Front fork legs - external spring type

1  Left-hand lower leg
2  Right-hand lower leg
3  Stanchion
4  O ring
5  Bush - 2 off
6  Sleeve nut - 2 off
7  Oil seal - 2 off
8  Lower spring seat
9  Fork spring
10  Upper spring seat - 2 off
11  * Spring shroud
12  Spacer - 2 off
13  Guide - 2 off
14  Lamp bracket
15  Lamp bracket
16  Seal - 2 off
17  Washer - 2 off
18  Top bolt - 2 off
19  Lower fork yoke
20  Bolt - 2 off
21  Guide
22  Pinch bolt
23  Plug - 2 off

* Note: late models have rubber gaiters

8.2a Check condition of seals and bushes

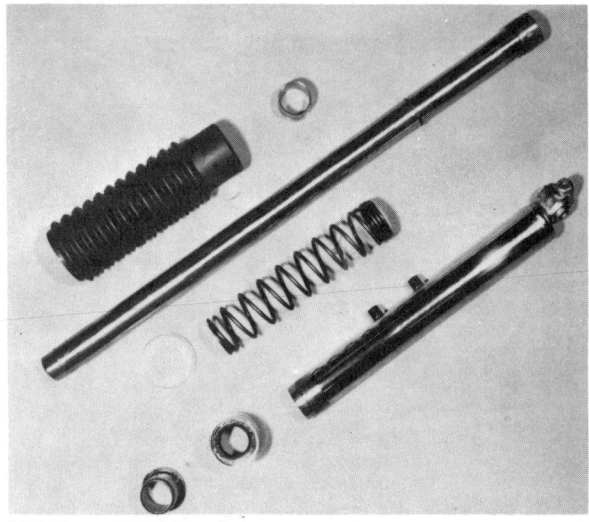

8.2b Drum brake type fork components

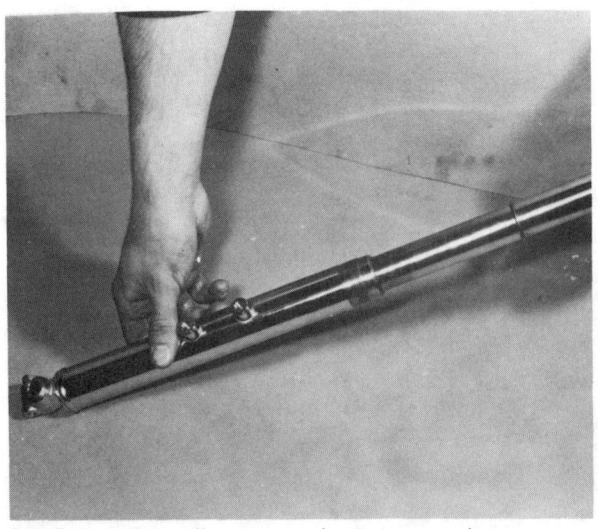

9.1a Follow dismantling sequence in reverse, ensuring ...

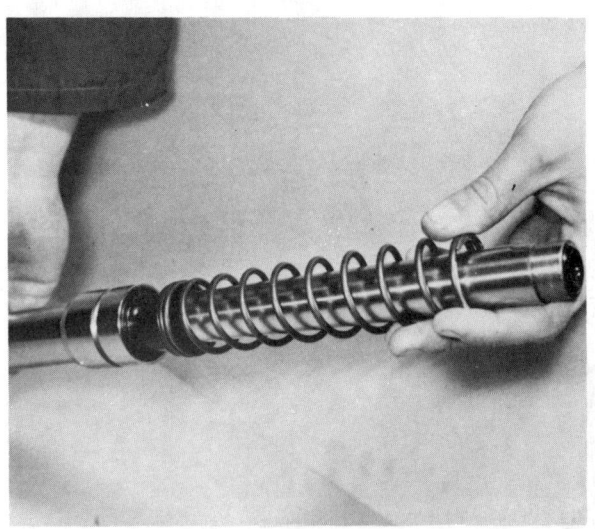

9.1b ... that spring seats correctly, and that ...

9.1c ... nylon washer and retainer are in place

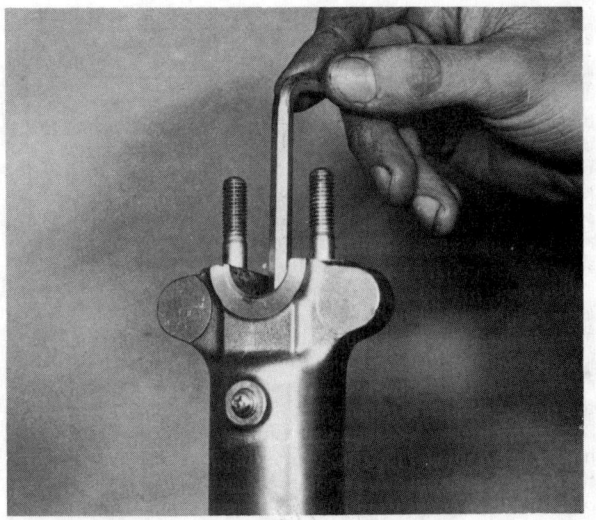

9.1d On disc brake type, tighten retaining screw

Chapter 4: Frame and forks

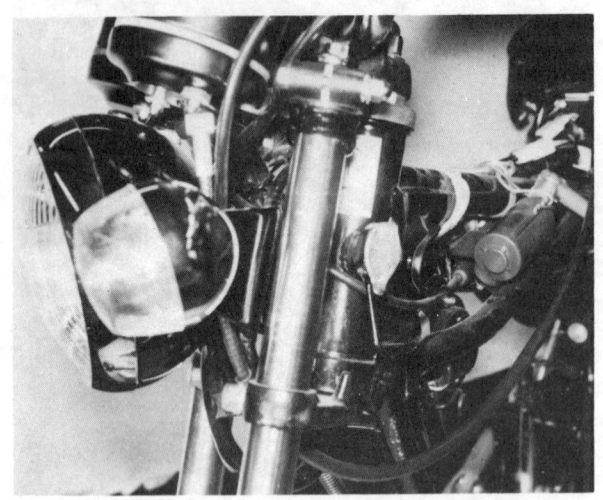

9.1e Refit assembled leg into yokes

9.1f Engine bolt can be used to draw leg into place

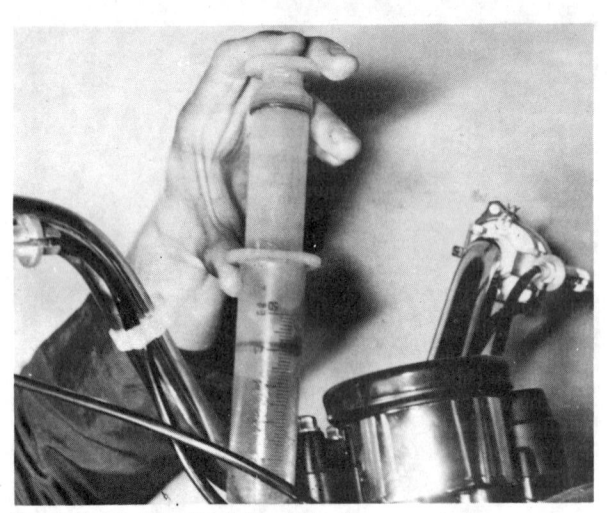

9.3a Fill each leg with 150 cc fork oil ...

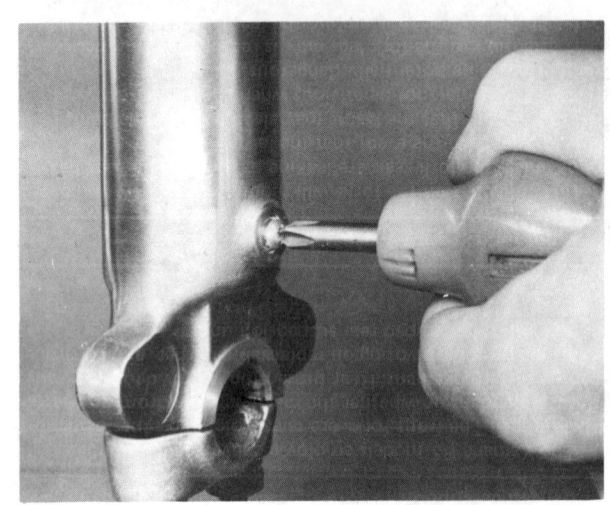

9.3b ... after tightening drain plugs!

**11 Frame: examination and renovation**

1  The frame is unlikely to require attention unless it is damaged as the result of an accident. In many cases, replacement of the frame is the only satisfactory course of action, if it is badly out of alignment. Comparatively few frame repair specialists have the necessary mandrels and jigs essential for the accurate re-setting of the frame and, even then there is no means of assessing to what extent the frame may have been overstressed such that a later fatigue failure may occur.
2  After a machine has covered an extensive mileage, it is advisable to keep a close watch for signs of cracking or splitting at any of the welded joints. Rust can cause weakness at these joints particularly if they are unpainted. Minor repairs can be effected by welding or brazing, depending on the extent of the damage found.
3  A frame out of alignment, will cause handling problems and may even promote 'speed wobbles' in a particular speed range. If misalignment is suspected as the result of an accident, it will be necessary to strip the machine so that the frame can be checked, and if needs be, renewed.

**12 Swinging arm rear fork: removal, examination and renovation**

1  The rear swinging arm fork pivots on two bushes, pressed into the fork crossmember. It is supported, on a long bolt or pivot shaft, between two webs which form the rear engine/gearbox mounting plates, and also a structural part of the frame. The unit is easily removed from the frame.
2  Place the machine on the centre stand, and secure it so that the rear wheel is raised clear of the ground. Remove the chainguard, which is retained by two screws on its outer edge, one at each end, and a third screw which is partially masked, on the inner edge.
3  Remove the final drive chain and place it to one side. It may be worthwhile relubricating the chain at this stage.
4  Detach the rear brake torque arm at the brake plate. It is held by a spring pin, nut and washer. Unscrew the rear brake adjusting nut and free the operating lever from the rod. Place the spring, trunnion and nut back on the rod, to avoid loss.
5  Remove the split pin and castellated nut from the left-hand end of the rear wheel spindle. Do not, at this stage, disturb the large, plain nut. Using a tommy bar, or suitable substitute,

## Chapter 4: Frame and forks

withdraw the wheel spindle from the right-hand side. The wheel will remain in position as it is retained by the outer, left-hand spindle. Pull out the distance piece between the brake plate and swinging arm, and also the right-hand chain adjuster. The brake plate assembly can now be removed, if desired, with the wheel and final drive assembly undisturbed.

6 Unscrew the large diameter nut on the outer spindle situated on the left-hand side. Disengage and remove the wheel complete with final drive sprocket and cush drive unit. The left-hand chain adjuster will now come away.

7 Unscrew the nut on the pivot shaft, leaving the shaft in position for the moment. Remove the two lower suspension unit nuts and bolt and raise the units clear of their mounting brackets. Withdraw the pivot shaft. It may be necessary to drive the shaft out using a long rod or bolt as a drift. Lift the swinging arm fork clear of the frame.

8 Examine the condition of the bushes in the fork crossmember. If damaged through wear or corrosion they should be renewed. If lateral play was evident with the fork in position they should also be renewed. The bushes can be drifted out of position, and new ones tapped in until they seat flush with the outer edge of the fork crossmember.

9 Check that the pivot shaft is not bent out of true, and clean any corrosion off before reassembly commences. It is advisable to coat the pivot shaft with grease to prevent corrosion taking place. Reassemble the swinging arm fork assembly by reversing the dismantling procedure. Tighten the pivot pin nut to 3 - 4.8 m kgs (21.7 - 34.7 lb f ft).

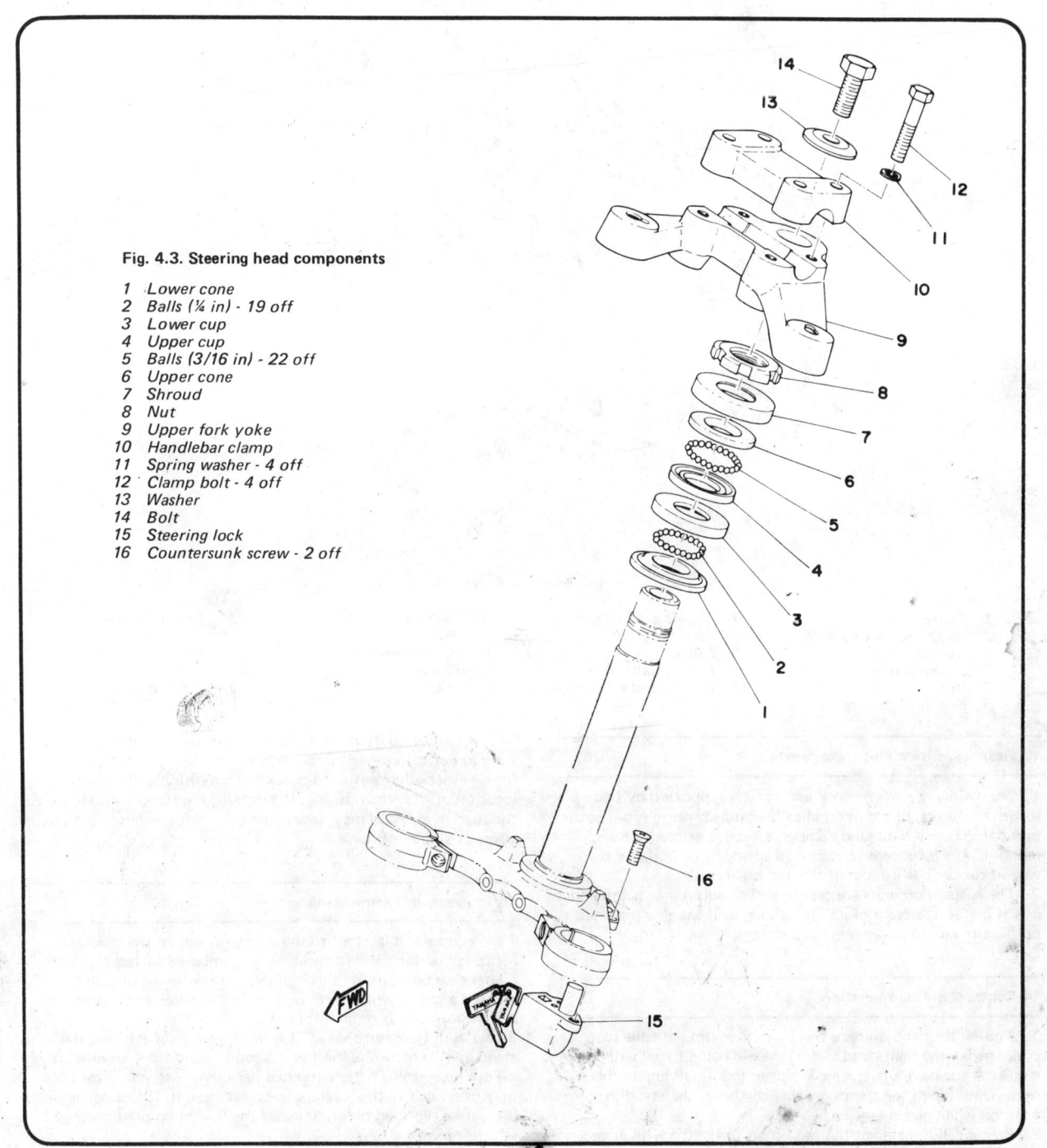

**Fig. 4.3. Steering head components**

1. Lower cone
2. Balls (¼ in) - 19 off
3. Lower cup
4. Upper cup
5. Balls (3/16 in) - 22 off
6. Upper cone
7. Shroud
8. Nut
9. Upper fork yoke
10. Handlebar clamp
11. Spring washer - 4 off
12. Clamp bolt - 4 off
13. Washer
14. Bolt
15. Steering lock
16. Countersunk screw - 2 off

Chapter 4: Frame and forks

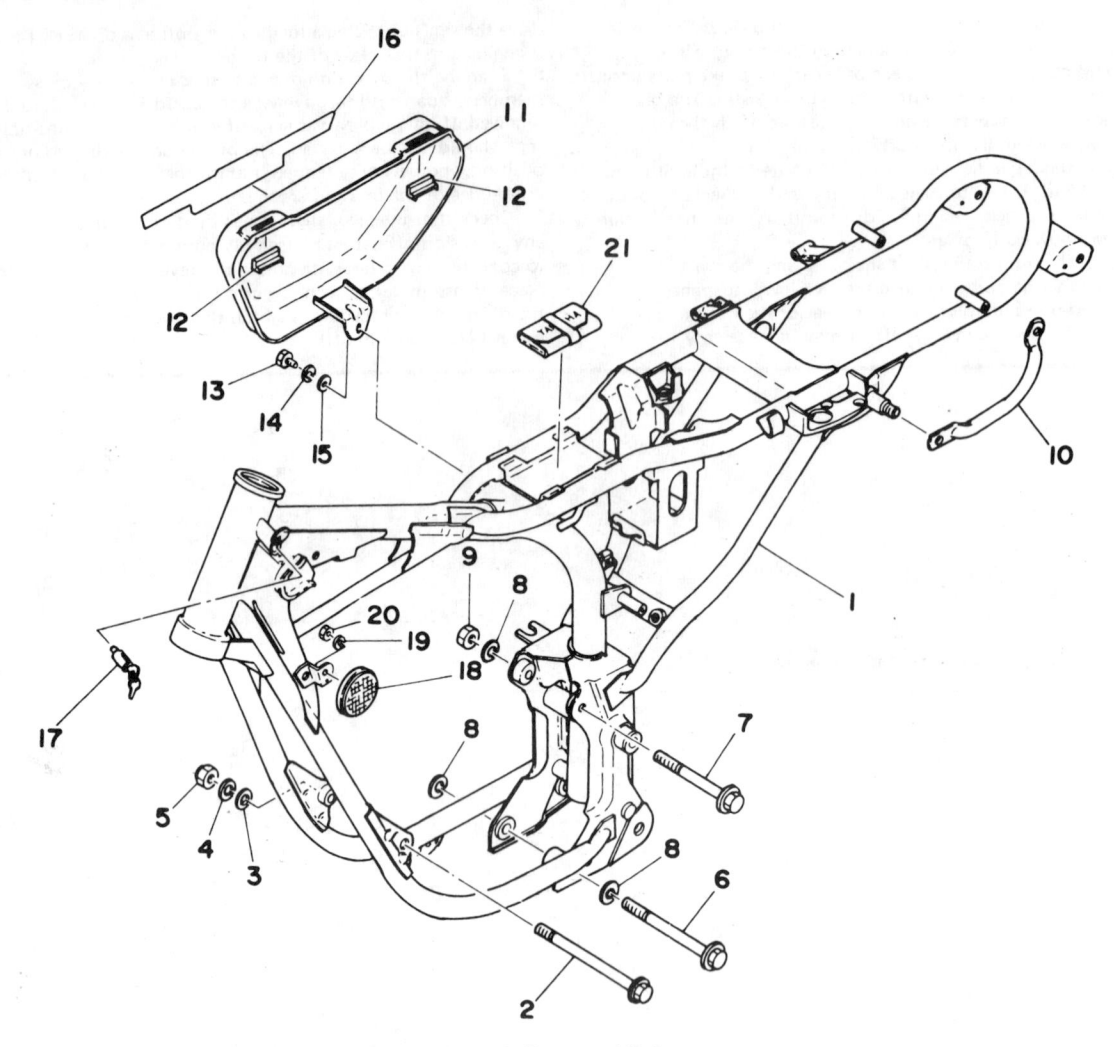

Fig. 4.4. Frame and fittings

| | | | |
|---|---|---|---|
| 1 Frame | 7 Engine mounting bolt | 12 Rubber | 17 Steering lock |
| 2 Engine mounting bolt | 8 Spring washer | 13 Bolt | 18 Reflector - 2 off |
| 3 Washer - 2 off | 9 Nut | 14 Spring washer | 19 Washer - 2 off |
| 4 Spring washer | 10 Lifting handle | 15 Washer | 20 Nut - 2 off |
| 5 Nut | 11 Right-hand side panel | 16 Decal | 21 Toolkit |
| 6 Engine mounting bolt | | | |

## 13 Rear suspension units: examination

1  The swinging arm rear fork assembly is supported by two suspension units, of the hydraulically damped spring type. Each unit consists of a hydraulic damper, effective primarily on rebound, and a concentric, chromed spring. It is mounted by way of rubber-bushed lugs at top and bottom.

2  The suspension units are sealed, and therefore no maintenance is feasible. In the event of a unit leaking, or if the damping fails, both units should be renewed as a matched pair.

## 14 Centre stand: examination

1  The centre stand shares a common pivot pin with the rear brake pedal, the right-hand end of the pin being fitted with a circlip. A strong return spring, attached to a small lug on the right-hand side of the stand, retracts and holds the stand in position when not in use.

2  Periodically, the pivot pin should be lubricated with grease. This is fairly important as its exposed position renders it susceptible to corrosion if left unattended.

3  Be especially careful to check the condition and correct location of the return spring. If this fails, the stand will fall onto the road in use, and may unseat the rider if it catches in a drain cover or similar obstacle.

## 15 Prop stand: examination

1  A prop stand is fitted to the left-hand side of the machine for quick parking and for parking on cambered surfaces. It pivots on a bolt passing through the footrest mounting plate, and has a spring which both holds it in the extended position, and also returns it when not in use.

2  As with the centre stand, it is important to check that the stand and its mounting bolt are in sound condition, and that it is kept lubricated. After extended periods of use, the pivot bolt may wear, due to the leverage imposed upon it. If this occurs, it must be renewed before it wears the mounting plate through which it passes.

71

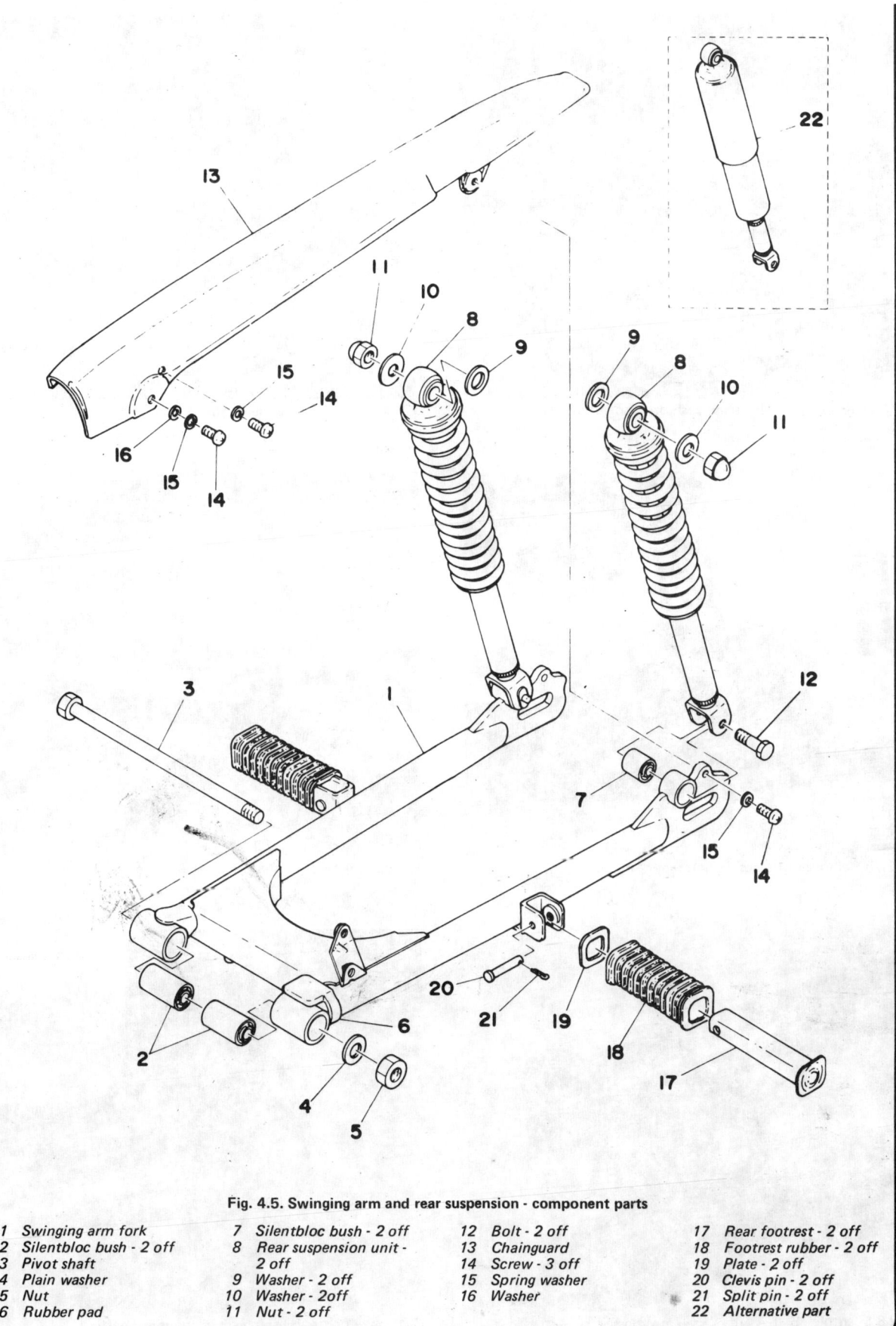

Fig. 4.5. Swinging arm and rear suspension - component parts

| 1 | Swinging arm fork | 7 | Silentbloc bush - 2 off | 12 | Bolt - 2 off | 17 | Rear footrest - 2 off |
| 2 | Silentbloc bush - 2 off | 8 | Rear suspension unit - 2 off | 13 | Chainguard | 18 | Footrest rubber - 2 off |
| 3 | Pivot shaft | | | 14 | Screw - 3 off | 19 | Plate - 2 off |
| 4 | Plain washer | 9 | Washer - 2 off | 15 | Spring washer | 20 | Clevis pin - 2 off |
| 5 | Nut | 10 | Washer - 2 off | 16 | Washer | 21 | Split pin - 2 off |
| 6 | Rubber pad | 11 | Nut - 2 off | | | 22 | Alternative part |

12.5 Remove rear wheel and cush drive/sprocket assembly

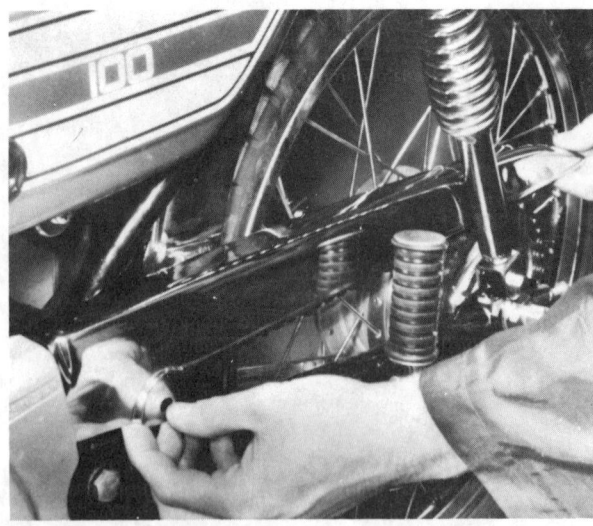

12.7a Chainguard can be removed to improve access

12.7b Remove lower suspension mounting bolts

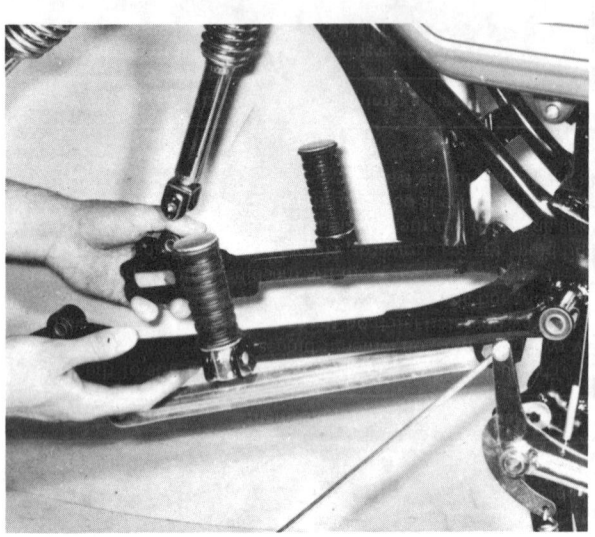

12.7c Displace pivot shaft and draw swinging arm clear

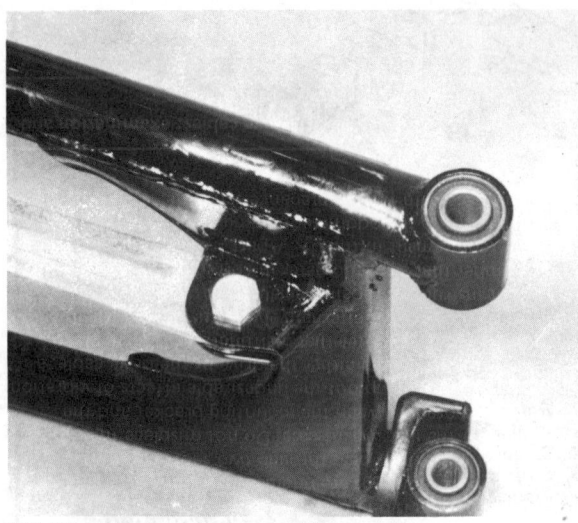

12.8 Silentbloc bushes are pressed into bosses

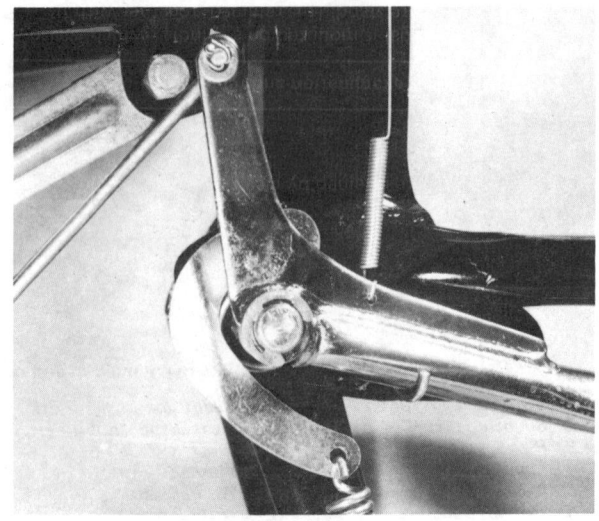

14.1 Brake pedal and stand share common shaft

Chapter 4: Frame and forks 73

## 16 Footrests: examination

1  The footrests comprise an assembly mounted below the frame and can be detached as a complete unit. They are rubber-mounted to damp out engine vibration. The assembly is retained by the lower rear engine bolt and one short bolt on the left-hand side. It is easier to remove the silencer, to facilitate removal.
2  Damage is likely only in the event of the machine being dropped. Slight bending can be rectified by stripping the assembly to the bare footrest bar, and straightening the bends by clamping the bar in a vice. A blowlamp should be applied to the affected area to avoid setting up stresses in the material, which may lead to subsequent fracturing.

16.1 Rear footrests are retained by clevis pin and split pin

## 17 Rear brake pedal: examination and renovation

1  The rear brake pedal is mounted on a spindle, which is also the pivot for the centre stand. It is retained, together with its return spring, by a single circlip on the right-hand side of the machine.
2  Should the pedal become bent in an accident, it can be straightened in a similar manner to that given for footrests in the preceding Section. If severely distorted, it should be renewed.

## 18 Dualseat: removal and replacement

1  The dualseat is attached to the right-hand side of the frame by two pivot pins, on which it hinges.
   A key operated latch on the left-hand side locks the seat in position, and also releases the helmet lock.
2  To remove the seat from the machine, unlock the seat latch and hinge the seat upwards. Remove the two R-shaped spring pins and push out the two small clevis pins on which the seat pivots. The hinge is bolted to the steel seat base, the other half being welded to the frame.

## 19 Kickstart lever: examination and renovation

1  The kickstart lever is splined and is secured to its shaft by means of a pinch bolt. The kickstart crank swivels so that it can be tucked out of the way when the engine is started. It is held in position on the swivel by a washer and circlip. A spring-loaded ball bearing locates the kickstart arm in either the operating or folded position; if the action becomes sloppy it is probable that the spring behind the ball bearing needs renewing. It is advisable to remove the circlip and washer occasionally, so that the kickstart crank can be detached and the swivel greased.
2  It is unlikely that the kickstart crank will bend in an accident unless the machine is ridden with the kickstart in the operating and not folded position. It should be removed and straightened, using the same technique as that recommended for the footrests in Section 16.2.

## 20 Speedometer and tachometer heads: removal and replacement

1  The Yamaha RS models are fitted with a centrally-mounted speedometer attached to the top fork yoke. On DX models, a tachometer is also fitted, adjacent to the speedometer. The speedometer and tachometer heads are rubber-mounted and attached to the top fork yoke by means of a single bolt fixing. If the bolt is slackened and the drive cable detached, the head complete with mounting can be lifted away; the yoke is slotted to aid removal. It will be necessary to remove the bulbs from the case of each instrument head by pulling the bulbholders from their seatings; each is retained by a rubber cup.
2  The rubber mountings are retained to each instrument case by two split pins which pass through two short columns attached to the base of the instrument case. Do not misplace the rubber cushion interposed between the mounting bracket and the instrument case to damp out the undesirable effects of vibration.
3  Apart from defects in either the drive or the drive cable, a speedometer or tachometer that malfunctions is difficult to repair. Fit a new one, or alternatively entrust the repair to a competent instrument repair specialist.
4  Remember that a speedometer in correct working order is a statutory requirement in the UK. Apart from this legal requirement, reference to the odometer reading is the best means of keeping in pace with the maintenance schedule.

## 21 Speedometer and tachometer drive cables: examination and maintenance

1  It is advisable to detach both cables from time to time in order to check whether they are lubricated adequately, and whether the outer coverings are compressed or damaged at any point along their run. Jerky or sluggish movements can often be attributed to a cable fault.
2  For greasing, withdraw the inner cable. After wiping off the

20.2 Instrument cases are retained by R-pin

old grease, clean with a petrol-soaked rag and examine the cable for broken strands or other damage.

3  Regrease the cable with high melting point grease, taking care not to grease the last six inches at the point where the cable enters the instrument head. If this precaution is not observed, grease will work into the head and immobilise the movement.

4  If either instrument ceases to function, suspect a broken cable. Inspection will show whether the inner cable has broken; if so, the inner cable alone can be renewed and reinserted in the outer casing, after greasing. Never fit a new inner cable alone if the outer covering is damaged or compressed at any point.

### 22 Speedometer and tachometer drive: location and examination

1  The speedometer drive gearbox is an integral part of the front wheel brake plate and is driven internally from the wheel hub. The gearbox rarely gives trouble if it is lubricated whenever the front wheel and brake plate are removed; there is no external grease nipple. If wear in the drive mechanism occurs, the worm complete with shaft can be withdrawn from the brake plate housing by unscrewing a pegged bush. The drive pinion is retained to the inside of the brake plate by a circlip, in front of the shaped driving plate that takes up the drive from the wheel hub. On disc brake machines, a separate drive gearbox is a push-fit on the hub, and can be pulled clear, after the wheel is detached.

2  The tachometer drive is taken from the gear pinion, integral with the clutch outer drum, via the kickstart idler pinion. If is unlikely that the drive will give trouble during normal service life of the machine.

### 23 Cleaning the machine

1  After removing all surface dirt with a rag or sponge washed frequently in clean water, the machine should be allowed to dry thoroughly. Application of car polish or wax to the cycle parts will give a good finish, particularly if the machine has been neglected for a long period.

2  The plated parts of the machine should require only a wipe with a damp rag. If the plated parts are badly corroded, as may occur during the winter when the roads are salted, it is preferable to use one of the proprietary chrome cleaners. These often have an oily base, which will help to prevent the corrosion from recurring.

3  If the engine parts are particularly oily, use a cleaning compound such as 'Gunk' or 'Jizer'. Apply the compound whilst the parts are dry and work it in with a brush so that it has the opportunity to penetrate the film of grease and oil. Finish off by washing down liberally with plenty of water, taking care that it does not enter the carburettor or the electrics. If desired, the now clean aluminium alloy parts can be enhanced further by using a special polish such as Solvol 'Autosol', which will fully restore their brilliance.

4  Whenever possible, the machine should be wiped down after it has been used in the wet, so that it is not garaged under damp conditions which will promote rusting. Make sure to wipe the chain and re-oil it, to prevent water from entering the rollers and causing harshness with an accompanying high rate of wear. Remember there is little chance of water entering the control cables and causing stiffness of operation if they are lubricated regularly as recommended in the Routine Maintenance Section.

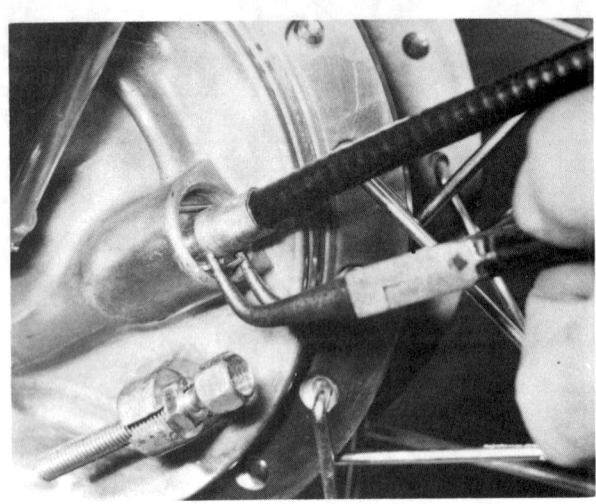

21.1 Drum brake models have cable retained by clip

22.1a Remove circlip, followed by ...

22.1b ... shim, to free driving dog

## Chapter 4: Frame and forks

22.1c Drive gear can now be pulled clear

22.1d Disc brake models have removable drive unit

### 24 Fault diagnosis: frame and forks

| Symptom | Cause | Remedy |
| --- | --- | --- |
| Machine veers either to the left or the right with hands off handlebars | Bent frame<br>Twisted forks<br>Wheels out of alignment | Check, and renew.<br>Check, and renew.<br>Check and realign. |
| Machine rolls at low speed | Overtight steering head bearings | Slacken until adjustment is correct. |
| Machine judders when front brake is applied | Slack steering head bearings | Tighten, until adjustment is correct. |
| Machine pitches on uneven surfaces | Ineffective fork dampers<br>Ineffective rear suspension units | Check oil content of front forks.<br>Check whether units still have damping action. |
| Fork action stiff | Fork legs out of alignment (twisted in yokes) | Slacken lower yoke clamps, and fork top bolts.<br>Pump fork several times then retighten from bottom upwards. |
| Machine wanders. Steering imprecise. Rear wheel tends to hop | Worn swinging arm pivot | Dismantle and renew bushes and pivot shaft. |

# Chapter 5 Wheels, brakes and tyres

*For information relating to 1977 on models refer to Chapter 7*

## Contents

| | |
|---|---|
| General description ... 1 | Wheel bearings: examination and replacement - disc brake models ... 13 |
| Front wheel: examination and renovation ... 2 | Wheel bearings: examination and replacement - front drum brake models, and rear wheel on all models ... 14 |
| Front wheel: removal, all models ... 3 | Speedometer drive gearbox: examination and lubrication ... 15 |
| Front drum brake: examination and renovation ... 4 | Rear wheel: examination and renovation ... 16 |
| Front disc brake: examination and renovation ... 5 | Rear wheel: removal and replacement ... 17 |
| Front disc brake: removing and replacing the disc and pads ... 6 | Adjusting the rear brake ... 18 |
| Front disc brake: removing, renovating and replacing the caliper unit ... 7 | Cush drive assembly: examination and renovation ... 19 |
| Front disc brake: master cylinder examination and renovation ... 8 | Rear wheel sprocket: removal, examination and replacement ... 20 |
| Front disc brake: hydraulic hose examination ... 9 | Final drive chain: examination and lubrication ... 21 |
| Front disc brake: bleeding the hydraulic system ... 10 | Chain and sprocket conversions ... 22 |
| Front disc brake: recommended torque settings ... 11 | Tyres: removal and replacement ... 23 |
| Front brake lever adjustment ... 12 | Fault diagnosis: wheels, brakes and tyres ... 24 |

## Specifications

**Tyres**
Front ... 2.75 x 18 in diameter
Rear ... 3.00 x 18 in diameter

**Tyre pressures**

| | Solo | Pillion or high speed |
|---|---|---|
| Front ... | 1.5 kgs/cm$^2$ (21 psi) | 1.8 kgs/cm$^2$ (26 psi) |
| Rear ... | 2.0 kgs/cm$^2$ (28 psi) | 2.3 kgs/cm$^2$ (33 psi) |

**Brakes**

| | RS100 and 125 | DX models |
|---|---|---|
| Front ... | Single leading shoe 112 mm or 152 mm | Hydraulic disc |
| Rear ... | Single leading shoe 112 mm or 132 mm | |

## 1 General description

1 On all Yamaha RS models, the front and rear wheels are of 18 inch diameter. The front rim carries a 2.75 inch section ribbed tyre, the rear rim being fitted with a block tread tyre of 3.00 inch section. The rims are of the chromium plated steel type, attached by spokes to cast aluminium hubs. On RS100 models, a single leading shoe front brake of 112 mm diameter, is integral with the hub, whilst on the later, DX models, the hub forms a mounting point for the stainless steel disc used in the hydraulic braking system. The rear wheel of all models is fitted with a 112 mm diameter drum brake of the single leading shoe type.

All wheels are of the quickly detachable type. The rear wheel can be removed without disturbing the rear wheel sprocket or drive chain.

## 2 Front wheel: examination and renovation

1 Place the machine on the centre stand so that the front wheel is raised clear of the ground. Spin the wheel and check the rim alignment. Small irregularities can be corrected by tightening the spokes in the affected area, although a certain amount of practice is necessary to prevent over-correction. Any flats in the wheel rim should be evident at the same time. These are more difficult to remove and in most cases it will be necessary to have the wheel rebuilt on a new rim. Apart from the effect on stability, a flat will expose the tyre bead and walls to greater risk of damage.

2 Check for loose or broken spokes. Tapping the spokes is the best guide to tension. A loose spoke will produce a quite different sound and should be tightened by turning the nipple in an

## Chapter 5: Wheels, brakes and tyres

anti-clockwise direction. Always re-check for run-out by spinning the wheel again. If the spokes have to be tightened an excessive amount, it is advisable to remove the tyre and tube by the procedure detailed in Section 23 of this Chapter; this is so that the protruding ends of the spokes can be ground off, to prevent them from chafing the inner tube and causing punctures.

### 3 Front wheel: removal, all models

1  With the machine supported on the centre stand, block the stand and crankcase to raise the wheel clear of the ground.
2  Detach the speedometer drive cable at the wheel, by unscrewing the knurled gland nut which retains it. On drum brake wheels, slacken the wheel spindle pinch bolt; disc brake models have a corresponding spindle clamp, which should be released by slackening the two bolts which retain it. Remove the front brake cable from drum brake models.
3  Remove the wheel spindle nut and split pin, and withdraw the spindle with the aid of a tommy bar. Take note of the wheel spacer fitted positions prior to removal, this will serve as a guide to correct reassembly. The wheel can now be lowered clear of the forks and put to one side.

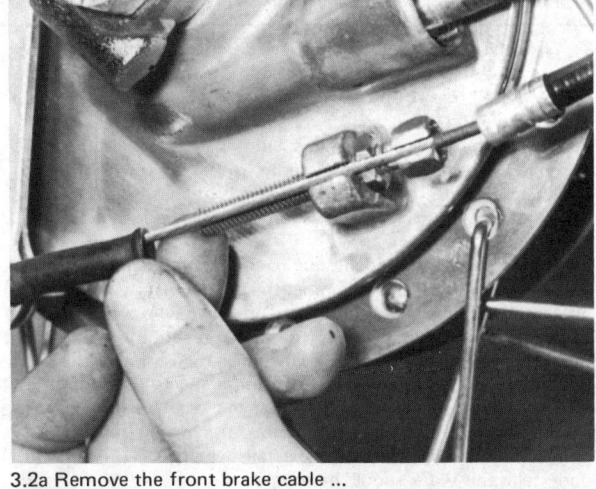

3.2a Remove the front brake cable ...

3.2b ... and speedometer drive cable

3.3a Slacken the wheel spindle nut, and ...

3.3b ... withdraw the spindle to release the front wheel

3.3c Note locating slot in speedometer drive (disc brake models)

Chapter 5: Wheels, brakes and tyres

## 4 Front drum brakes: examination and renovation

1 With the front wheel removed, as described in the preceding section, this single leading shoe brake mechanism and backplate can be pulled free from the drum.
2 Examine the drum surface for signs of scoring or oil contamination. Both of these conditions will impair braking efficiency. Remove all traces of dust, preferably using a brass wire brush, taking care not to inhale any of it, as it is of an asbestos nature, and consequently toxic. Remove oil or grease deposits, using a petrol soaked rag.
3 If deep scoring is evident, due to the linings having worn through to the shoe at some time, the drum must be skimmed on a lathe, or renewed. Whilst there are firms who will undertake to skim a drum whilst fitted to the wheel, it should be borne in mind that excessive skimming will change the radius of the drum in relation to the brake shoes, therefore reducing the friction area until extensive bedding in has taken place. Also full adjustment of the shoes may not be possible. If in doubt about this point, the advice of one of the specialist engineering firms who undertake this work should be sought.
4 If fork oil or grease from the wheel bearings has badly contaminated the linings, they should be renewed. There is no satisfactory way of degreasing the lining material, which in any case is relatively cheap to replace. It is a false economy to try to cut corners with brake components; the whole safety of both machine and rider being dependent on their condition.
5 The linings are bonded to the shoes, and the shoes must be renewed complete with the new linings. Renewal is accomplished by folding the shoes together until the spring tension is relaxed, and then lifting the shoes and springs off the brake plate. Fitting new shoes is a direct reversal of the above procedure.
Should the shoes appear worn, this can be checked by measuring the diameter of the lining material with the shoes in place on the brake plate. If the total diameter is 5 mm smaller than the relevant figure quoted below, the shoes should be renewed as a pair.

**Brake linings - nominal diameters:**

|  | RS100 | RS125 |
|---|---|---|
| Front | 110 mm (4.33 in) | 150 mm (5.90 in) |
| Rear | 110 mm (4.33 in) | 130 mm (5.12 in) |
| Overall wear limit: | 5 mm (0.20 in) | |

6 Before refitting existing shoes, roughen the lining surface sufficiently to break the glaze which will have formed in use.
7 Push out the fulcrum from the brake plate. If there is corrosion on the fulcrum face or in its bore, this should be removed using wet or dry paper. Grease the fulcrum before installation.

## 5 Front disc brake: examination and renovation

1 The introduction of the DX or disc brake models reflects the growing popularity of this type of brake. It provides the maximum braking power from a compact and lightweight unit, and has the advantage of ease of maintenance. The stainless steel disc, mounted on the right-hand side of the hub, is of maximum cosmetic appeal. The caliper, mounted forward of the front right-hand fork leg, is of the floating type. That is, when the single piston bears upon the disc, the caliper unit slides fractionally to the right, to bring the second, fixed, pad into use. This provides adequate pressure on the disc surfaces, without the added complexity of a double piston caliper. Hydraulic pressure is fed to the caliper by way of armoured hoses and hydraulic pipes. The combined master cylinder and reservoir unit is mounted on the right-hand side of the handlebar, the operating lever forming an integral part of the unit, and acting directly on the master cylinder piston.
2 Pad removal and replacement can be undertaken with the wheel removed, or alternatively, by removing the caliper unit from the fork leg. In each case, it is a simple operation, with neither method being easier nor quicker than the other.

4.1 Front brake plate assembly can be lifted out

4.2 Examine drum surface for scoring

4.4 Check linings for contamination and wear

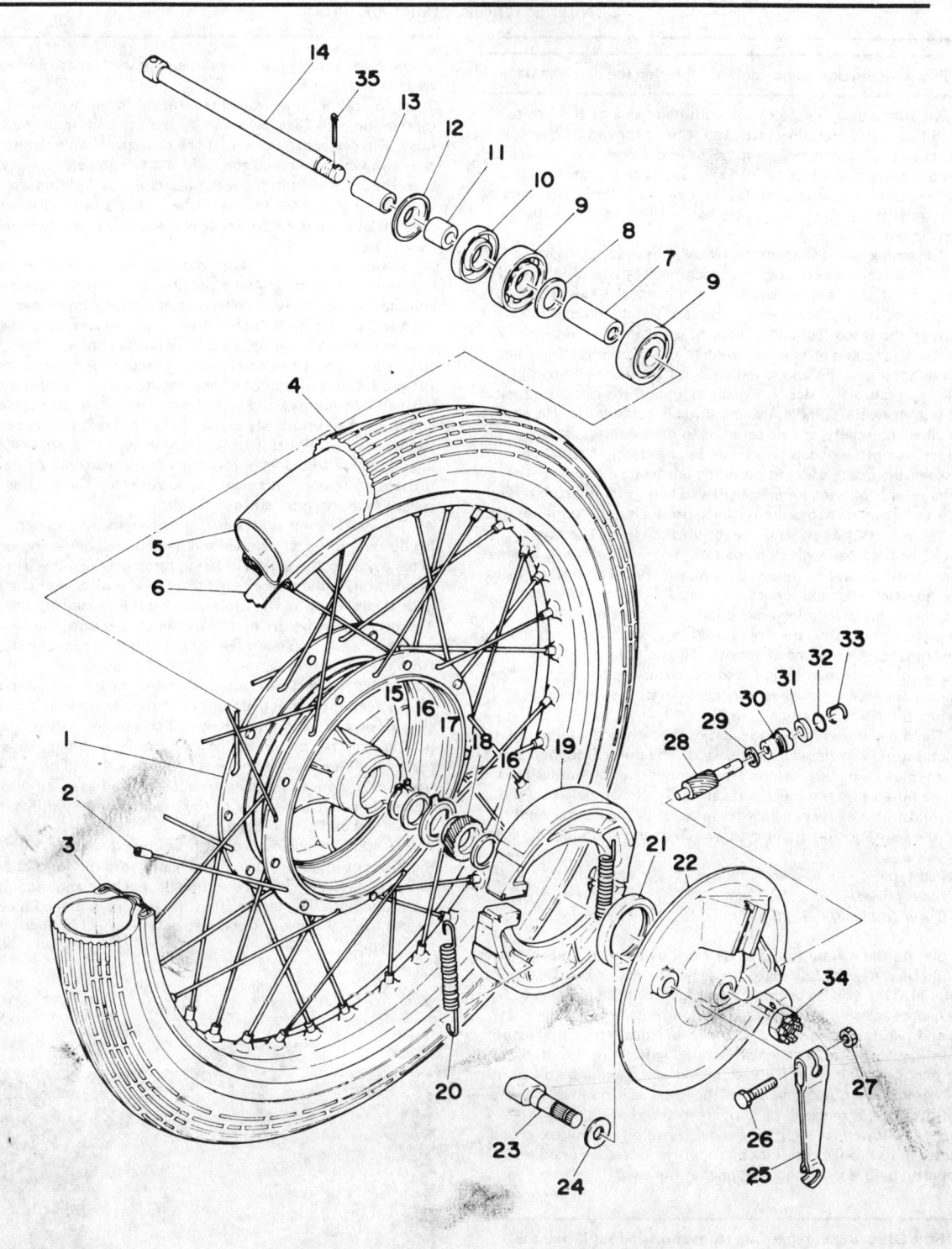

Fig. 5.1. Front wheel - drum brake models

| | | | |
|---|---|---|---|
| 1 Hub | 10 Seal | 19 Brake shoe - 2 off | 28 Driven gear |
| 2 Spoke set | 11 Distance piece | 20 Spring - 2 off | 29 Washer |
| 3 Wheel rim - | 12 Cover | 21 Oil seal | 30 Body |
| 4 Tyre | 13 Sleeve | 22 Brake plate | 31 Oil seal |
| 5 Inner tube | 14 Wheel spindle | 23 Fulcrum | 32 O ring |
| 6 Rim tape | 15 Circlip | 24 Washer | 33 Clip |
| 7 Bearing spacer | 16 Washer - 2 off | 25 Actuating lever | 34 Castellated nut |
| 8 Washer | 17 Driving dog | 26 Pinch bolt | 35 Split pin |
| 9 Wheel bearing - 2 off | 18 Driving gear | 27 Nut | |

## 6 Front disc brake: removing and replacing the disc and pads

1 The brake disc, bolted to the right-hand side of the front wheel hub, rarely requires attention. Check for run out, which may have occurred as the result of crash damage, and for wear. Run out should not exceed 0.15 mm at any point, and the disc itself must not be permitted to wear below the limit thickness of 4.6 mm. If these figures are exceeded in either case, the disc must be renewed.

2 To remove the disc from the wheel, it is first necessary to detach the front wheel from the machine. The machine should be placed on the centre stand so that the wheel is raised clear of the ground. It may be necessary to place thin wooden blocks between the stand feet and ground to assist in this. Detach the speedometer cable after unscrewing the knurled retaining ring. Remove the split pin and castellated nut from the end of the wheel spindle. After slackening the clamp on the opposite fork leg, withdraw the spindle, using a suitable tommy bar. Remove the wheel complete, taking care not to damage the disc or caliper. A small wooden wedge can be inserted in the caliper between the brake pads, to prevent their being expelled if the brake lever is operated inadvertently. There is in any case a tendency to 'creep' in hydraulic systems, producing the same results.

3 The disc itself is attached directly to the hub. The four mounting bolts are locked in position by two double tab washers, which must always be renewed on reassembly. The disc surfaces may now be examined for scoring, which is usually caused by particles of grit becoming embedded in the friction material of the pads. If excessive, the disc should be renewed as deep scoring is detrimental to braking efficiency. Reassembly is a direct reversal of the dismantling procedure, ensuring that each component is thoroughly cleaned, especially mating surfaces, to obviate the risk of misalignment.

4 The inner (fixed) pad is located by a screw and support plate, which should be removed to release it. The outer (moving) pad can then be eased out, using a small screwdriver or similar tool. It will be noted that the inner pad can easily be identified by the small locating peg, which engages in a slot in the caliper, and also the hole for the retaining screw. Examine the pads for wear:

| Pad type | Nominal size | Wear limit |
| --- | --- | --- |
| Inner (fixed) | 12.0 mm | 7.5 mm |
| Outer (moving) | 15.5 mm | 11.0 mm |

Should the pad be found to be worn below the limits given, it should be renewed. Look also for signs of staining on the friction material. This may be caused by leakage from the fork leg or from the caliper seals; in either case attention must be given to locating and rectifying the source of the leak. Ensure that the piston is not inadvertently expelled whilst the pads are removed. Should this occur, air will enter the system and will have to be bled out. Reassembly is a direct reversal of the removal sequence. Be sure that the pads, if re-used, are kept clean and that no foreign matter is allowed to enter the caliper. Ensure that the fixed pad locates in its groove correctly and that the brake is functioning properly, before using the machine on the road.

## 7 Front disc brake: removing, renovation and replacing the caliper unit

1 The caliper unit is of the floating type. That is, the caliper body has a controlled sliding motion to allow the fixed pad to be brought into contact with the disc, under pressure, from the opposing, moving pad. This movement is effected by allowing the body to slide, via two pins, in relation to the mounting bracket. The mounting bracket and body can be separated after removing two circlips, threading a 5 mm screw into the tapped end of the pin, and withdrawing by pulling on the head of the screw with pliers. The two parts, however, cannot be obtained separately, so it is better to assume that if play has developed in the sliding mount, then the assembly will be in need of renewal in any case. Judder under braking may well be attributable to wear at this point.

2 If the caliper unit warrants removal for inspection or renovation, it is first necessary to remove and drain the hydraulic hose. Disconnect the union at the caliper. Have a suitable container in which to catch the fluid. At this stage, it is as well to stop the flow of fluid from the reservoir, by holding the front brake lever in against the handlebar. This is easily done using a stout elastic band, or alternatively, a section cut from an old inner tube.

3 **Note:** Brake fluid will discolour or remove paint if contact is allowed. Avoid this where possible and remove accidental splashes immediately. Similarly, avoid contact between the fluid and plastic parts such as instrument lenses, as damage will also be done to these. When all the fluid is drained from the hose, clean the connections carefully and secure the hose end and fittings inside a clean polythene bag, to await reassembly. As with all hydraulic systems, it is most important to keep each component scrupulously clean, and to prevent the ingress of any foreign matter. For this reason, it is as well to prepare a clean area in which to work, before further dismantling. As in any form of component dismantling, ensure that the outside of the caliper is thoroughly cleaned down.

4 The caliper unit is attached to the inside of the right-hand fork leg by two bolts, which, when removed, will allow the unit to be lifted away. If the caliper is being removed with the front wheel in position, it should be lifted clear of the disc. Remove the brake pads as described in the preceding Section, exposing the piston. The piston may be driven out of the caliper body by an air jet - a foot pump if necessary. Remove the piston seal and dust seal from the caliper body.

5 Clean each part carefully, using only clean hydraulic fluid. On no account use petrol, oil or paraffin as these will cause the seals to degrade and swell. Keep all components dust free.

6 Examine the piston surface for scoring or pitting, any imperfection will necessitate renewal. The seals should be renewed as a matter of course; re-using an old seal is a false economy. Remember that the safety of the machine is very much dependent on seal and piston condition.

7 Reassemble, again ensuring absolute cleanliness, by reversing the dismantling procedure. Use clean hydraulic fluid as a lubricant. Replace the caliper unit on the machine and reconnect the hydraulic hose. Remember that the system will need bleeding before use, by following the instructions given in Section 10 of this Chapter.

6.4a Remove screw and retainer, releasing the ...

**Fig. 5.2. Front disc brake - hydraulic components**

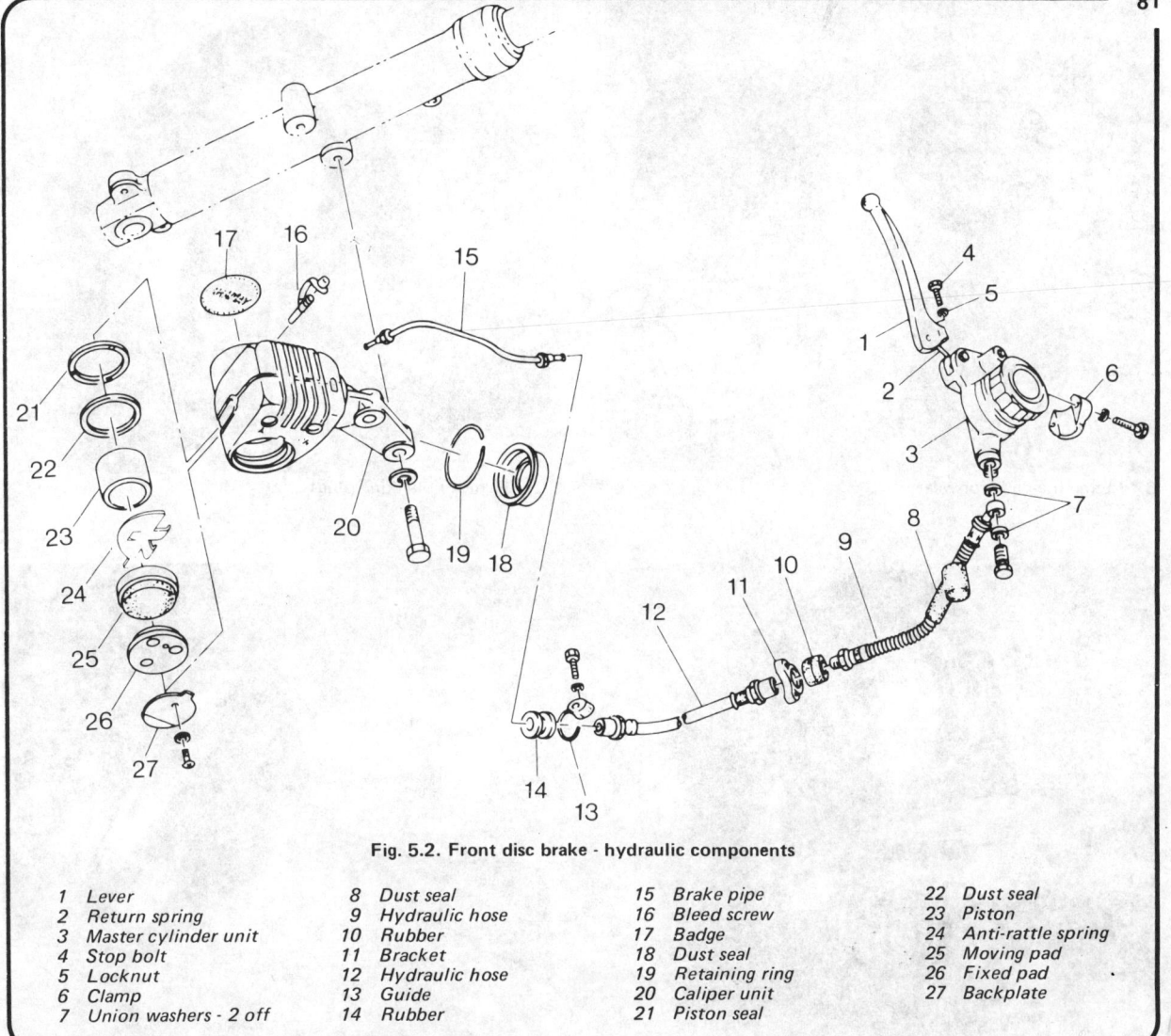

| | | | |
|---|---|---|---|
| 1 Lever | 8 Dust seal | 15 Brake pipe | 22 Dust seal |
| 2 Return spring | 9 Hydraulic hose | 16 Bleed screw | 23 Piston |
| 3 Master cylinder unit | 10 Rubber | 17 Badge | 24 Anti-rattle spring |
| 4 Stop bolt | 11 Bracket | 18 Dust seal | 25 Moving pad |
| 5 Locknut | 12 Hydraulic hose | 19 Retaining ring | 26 Fixed pad |
| 6 Clamp | 13 Guide | 20 Caliper unit | 27 Backplate |
| 7 Union washers - 2 off | 14 Rubber | 21 Piston seal | |

6.4b ... fixed friction pad

6.4c Inner (moving) pad can be displaced. Note pin

6.4d Examine pads for wear

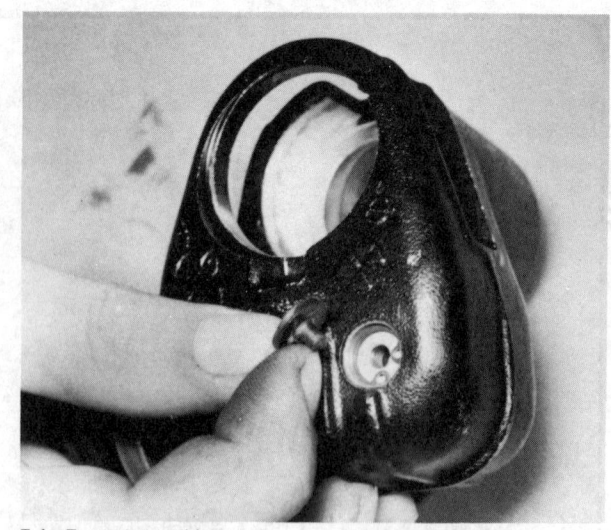

7.1a Remove blanking plugs ...

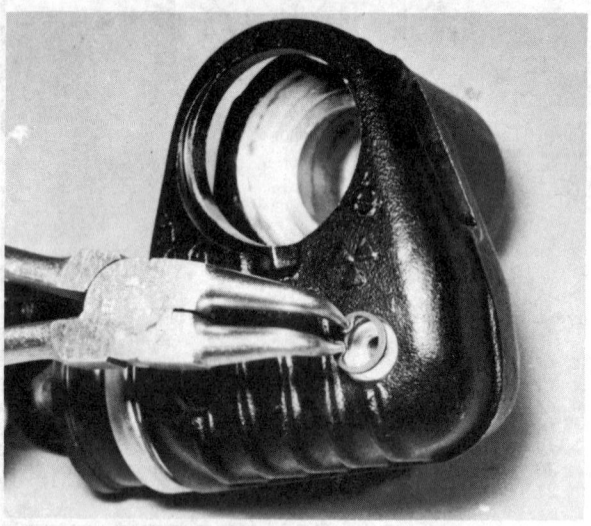

7.1b ... and circlips

7.1c Pull out mounting pins with bolt and pliers

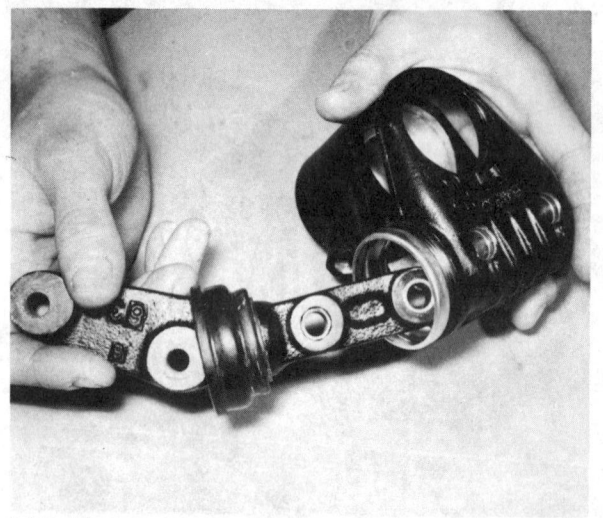

7.1d Mounting bracket can now be withdrawn

7.1e ... as can the anti-rattle spring

# Chapter 5: Wheels, brakes and tyres

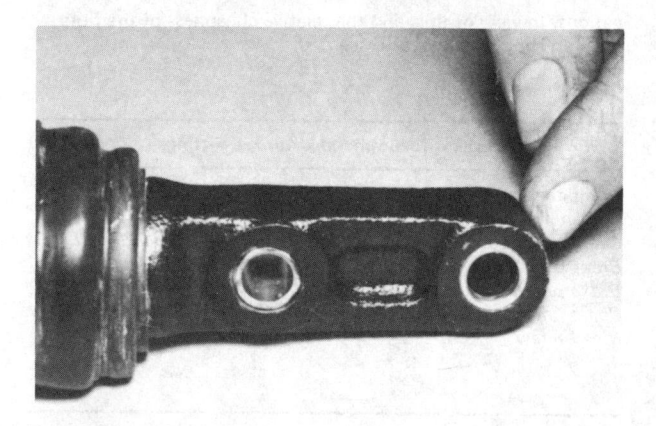

7.1f Wear in bushes can cause judder

7.4a Caliper is bolted to fork leg

7.4b Compressed air can be used to displace piston

7.4c Seals seat in grooves in body

## 8 Front disc brake: master cylinder examination and renovation

1  The master cylinder forms a unit with the hydraulic fluid reservoir and front brake lever, and is mounted by a clamp to the right-hand side of the handlebars.
2  The unit must be drained before any dismantling can be undertaken. Place a suitable container below the caliper unit and run a length of plastic tubing from the caliper bleed screw to the container. Unscrew the bleed screw one full turn and proceed to empty the system by squeezing the front brake lever. When all the fluid has been expelled, tighten the bleed screw and remove the tube.
3  Select a suitable clean area in which the various components may be safely laid out, a large piece of white lint-free cloth or white paper being ideal.
4  Remove the front brake lever pivot bolt, and lift away the lever, and also the front brake lamp switch located adjacent to the lever. Detach the hose by undoing the banjo mounting at the end of the cylinder. Remove the screwed master cylinder cap, or the tamper proof cover plate, depending on type fitted, and withdraw the diaphragm. Remove the two screws holding the unit to the handlebars, and lift away. Pull off the rubber boot fitted to the end of the cylinder, exposing the circlip. Remove the circlip using circlip pliers to reach it, and withdraw the piston assembly and spring.
5  Examine the piston and seals for scoring or wear and renew if imperfect. Excessive scoring may be due to contaminated fluid, and if this is suspected, it is probably worth checking the condition of the caliper seals and piston.
6  Reassemble carefully, using hydraulic fluid as a lubricant on seals and piston, reversing the dismantling sequence. Make sure the rubber boot is fitted correctly, and that the unit is clamped securely to the handlebars. Reconnect the hydraulic hose, tightening the banjo union bolt to the recommended torque setting. Refill the reservoir, remembering to top up after the system has been bled by following the procedure given in Section 10 of this Chapter.

## Chapter 5: Wheels, brakes and tyres

2 The hose, of the flexible armoured type, must withstand considerable pressure in service, and whilst it is easily ignored, it should be checked carefully as a sudden failure can be potentially fatal. Look not only for signs of chafing against the wheel or fork leg, but also for any stains due to fluid seeping from cracks in the hose or from the connections at either end.

### 10 Front disc brake: bleeding the hydraulic system

1 Removal of all the air from the hydraulic system is essential to the efficiency of the braking system. Air can enter the system due to leaks or when any part of the system has been dismantled for repair or overhaul. Topping the system up will not suffice, as air pockets will still remain, even small amounts causing dramatic loss of brake pressure.
2 Check the level in the reservoir, and fill almost to the top. Again, beware of spilling the fluid on to painted or plastic surfaces.
3 Place a clean jar below the brake caliper unit and attach a clear plastic tube from the caliper bleed screw to the container. Place some clean hydraulic fluid in the container so that the pipe is always immersed below the surface of the fluid.
4 Unscrew the bleed screw one complete turn and pump the handlebar lever slowly. As the fluid is ejected from the bleed screw the level in the reservoir will fall. Take care that the level does not drop too low whilst the operation continues, otherwise air will re-enter the system, necessitating a fresh start.
5 Continue the pumping action with the lever until no further air bubbles emerge from the end of the plastic pipe. Hold the brake lever against the handlebar and tighten the caliper bleed screw. Remove the plastic tube **after** the bleed screw is closed.
6 Check the brake action for sponginess, which usually denotes there is still air in the system. If the action is spongy, continue the bleeding operation in the same manner, until all traces of air are removed.
7 When all traces of air have been removed from the system, top up the reservoir and refit the diaphragm and cap or cover, as appropriate. Check the entire system for leaks, and check also that the brake system in general is functioning efficiently before using the machine on the road.
8 Brake fluid drained from the system will almost certainly be contaminated, either by foreign matter or more commonly by the absorption of water from the air. All hydraulic fluids are to some degree hygroscopic, that is, they are capable of drawing water from the atmosphere, and thereby degrading their specifications. In view of this, and the relative cheapness of the fluid, oil fluid should always be discarded.

### 11 Front disc brake: recommended torque settings

| Component | kg/cm | Imperial |
| --- | --- | --- |
| Disc mounting bolts | 200 - 250 | (14.5 - 18 ft lbs) |
| Caliper mounting bolts | 230 - 280 | (17.0 - 20 ft lbs) |
| Brake pipe union nut | 130 - 180 | (9.4 - 13 ft lbs) |
| Brake pipe/hose union nut | 150 - 200 | (11.0 - 14.5 ft lbs) |
| Banjo union nut at brake lever | 230 - 280 | (17.0 - 20 ft lbs) |
| Bleed screw | 50 - 70 | (3.6 to 5.0 ft lbs) |

### 12 Front brake lever adjustment

The front brake lever should be checked periodically and adjusted at the stop bolt to give 13 mm - 20 mm (0.5 - 0.8 in) free play measured at the lever end.

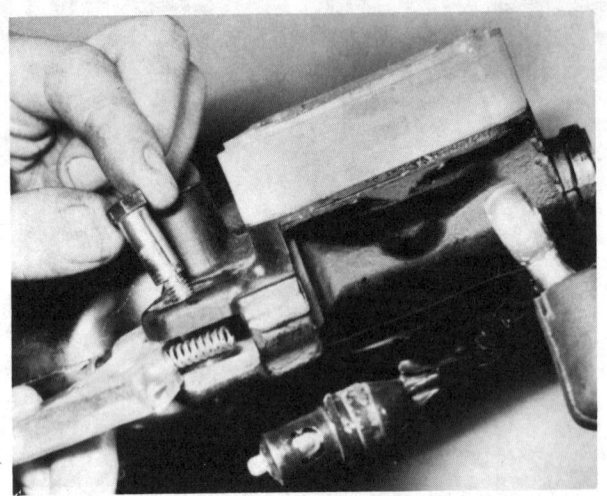

8.4a Remove lever and switch unit ...

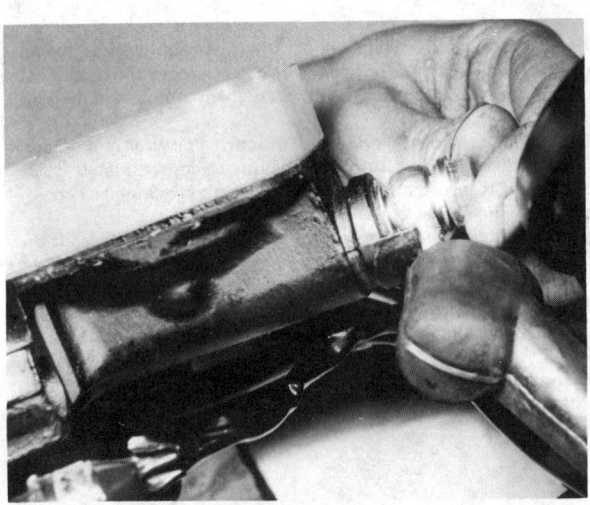

8.4b ... and banjo union

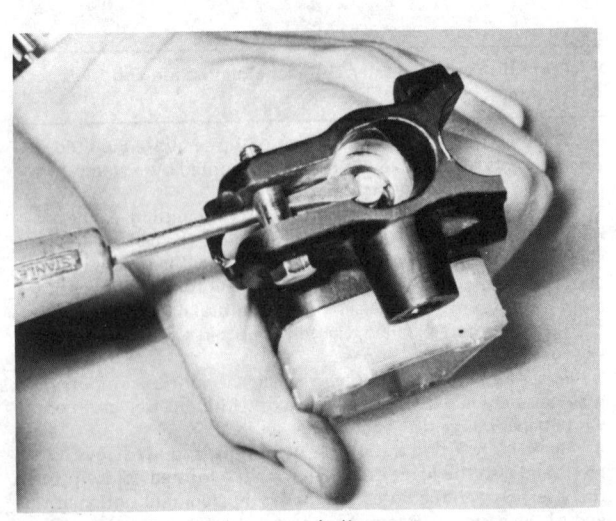

8.4c Depress piston and remove circlip

## Chapter 5: Wheels, brakes and tyres

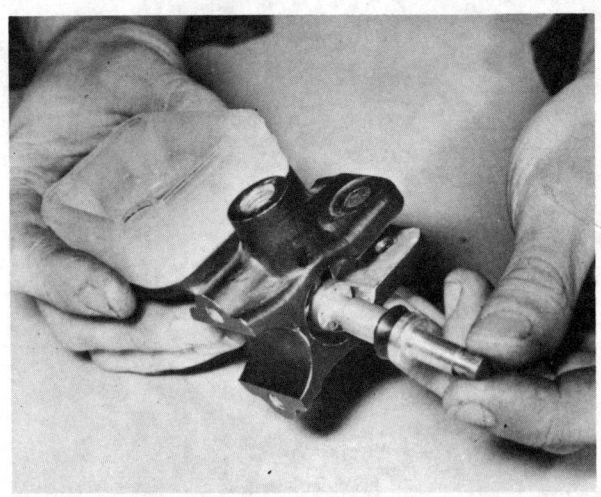

8.4d Piston can be withdrawn along with ...

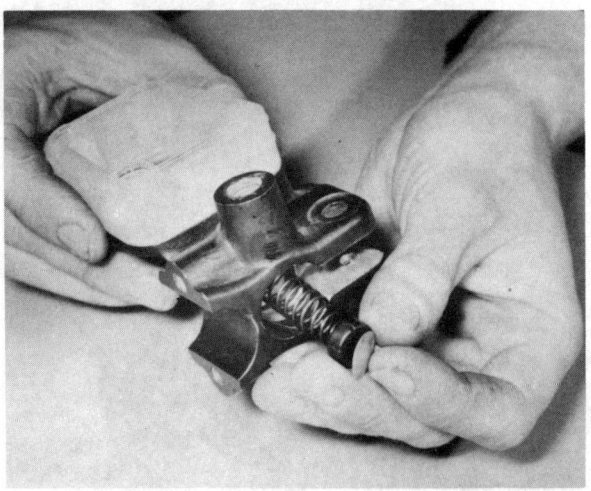

8.4e ...spring and seat

8.5a Examine components for wear or damage

8.5b Reservoir may have screwed cap or cover (above)

### 13 Wheel bearings: examination and replacement - disc brake models

1  Access to the front wheel bearings may be made after removal of the wheel from the forks. Pull the speedometer gearbox out of the hub left-hand boss and remove the dust seal cover and wheel spacer from the hub right-hand side.
2  Lay the wheel on the ground with the disc side facing downward and with a special tool, in the form of a rod with a curved end, insert the curved end into the hole in the centre of the spacer separating the two wheel bearings. If the other end of the special tool is hit with a hammer, the right-hand bearing, bearing flange washer, and bearing spacer will be expelled from the hub.
3  Invert the wheel and drive out the left-hand bearing by inserting a drift of the appropriate size, through the hub. During the removal of either bearing it may be necessary to support the wheel across an open-ended box so that there is sufficient clearance for the bearing to be displaced completely from the hub.
4  Remove all the old grease from the hub and bearings, giving the latter a final wash in petrol. Check the bearings for signs of play or roughness when they are turned. If there is any doubt about the condition of a bearing, it should be renewed.

5  Before replacing the bearings, first pack the hub with new grease. Then drive the bearings back into position, not forgetting the distance piece that separates them. Take great care to ensure that the bearings enter the housings perfectly squarely otherwise the housing surface may be broached. Fit replacement oil seals and any dust covers or spacers that were also displaced during the original dismantling operation.

### 14 Wheel bearings: examination and replacement - front drum brake models, and rear wheel on all models

1  The sequence given in the preceding Section is also applicable to models fitted with front drum brakes. Rear wheel bearings on all models can be dealt with in a similar manner.

### 15 Speedometer drive gearbox: examination and lubrication

1  The speedometer drive is taken from a gearbox mounted on the front wheel of all models. On drum brake models, the drive gear is supported in the brake plate casting, as is the driven shaft.

Chapter 5: Wheels, brakes and tyres

14.1a Cush drive unit pulls out of wheel

14.1b Bearings are driven into hubs and cush drive

14.1c Note spacer over outer shaft ...

14.1d ...and headed spacer in hub

The drive gear can be removed from the inside of the brake plate, together with the driving dog, after removing the circlip which retains it. The driven shaft can be removed from the outside of the brake plate by unscrewing its retaining gland nut. Unless excessively worn due to lack of lubricant, which will necessitate renewal, little can be done by way of maintenance other than thorough cleaning and lubrication of the components and their housing.

2  Maintenance of the disc brake type of gearbox should be approached in the same manner; in this case, the unit being a self-contained assembly which is a push fit in the front hub. The unit should be greased as a matter of course, each time the front wheel is removed.

### 16  Rear wheel: examination and renovation

1  The procedure given in Section 2 of this Chapter is also applicable to the rear wheels of all RS models.

### 17  Rear wheel: removal and replacement

1  It is not necessary to remove the final drive chain in order to remove the rear wheel. Place the machine on its centre stand. Unscrew the rear brake adjuster and release the brake rod. The spring and adjuster should be replaced on the rod for safe keeping. Remove the spring pin which retains the torque arm nut, and remove the nut and torque arm.

2  Remove the split pin and castellated nut from the left-hand end of the rear wheel spindle. Do not disturb the larger, plain nut. Using a tommy bar, or suitable substitute, withdraw the wheel spindle from the right-hand side of the wheel. The wheel will remain in position as it is retained by the outer spindle, from the left-hand side. The distance piece between the brake plate and swinging arm can now be pulled clear. If attention to the brake only is required, this can be removed with the wheel in position.

3  The wheel may be removed easily, leaving the final drive sprocket and chain in position, by pulling the wheel to the right, which will separate the cush drive assembly.

4  Reassembly is a direct reversal of the above procedure, ensuring that the rubber cush drive segments are positioned correctly. Although undisturbed it is as well to check the tension of the final drive chain at this stage. Readjust the rear brake.

## Tyre changing sequence — tubed tyres

**A** — Deflate tyre. After pushing tyre beads away from rim flanges push tyre bead into well of rim at point opposite valve. Insert tyre lever adjacent to valve and work bead over edge of rim.

**B** — Use two levers to work bead over edge of rim. Note use of rim protectors.

**C** — Remove inner tube from tyre.

**D** — When first bead is clear, remove tyre as shown.

**E** — When fitting, partially inflate inner tube and insert in tyre.

**F** — Work first bead over rim and feed valve through hole in rim. Partially screw on retaining nut to hold valve in place.

**G** — Check that inner tube is positioned correctly and work second bead over rim using tyre levers. Start at a point opposite valve.

**H** — Work final area of bead over rim whilst pushing valve inwards to ensure that inner tube is not trapped.

Chapter 5: Wheels, brakes and tyres

17.1 Disconnect brake rod and torque arm

17.2a Slacken spindle nut, and ...

17.2b ... withdraw spindle to free wheel

## 18 Adjusting the rear brake

1  If the adjustment of the rear brake is correct, the rear brake pedal will have about 25 mm (1 inch) free play before the brake commences to operate.
2  The length of travel is controlled by the adjuster at the end of the brake operating rod, close to the brake operating arm. If the nut is turned clockwise, the amount of travel is reduced and vice-versa. Always check that the brake is not binding after adjustments have been made.
3  Note that it may be necessary to re-adjust the height of the stop lamp switch if the pedal height has been changed to any marked extent. The switch is located immediately below the right-hand side cover that carries the capacity symbol of the model. The body of the switch is threaded, so that it can be raised or lowered, after the locknuts have been slackened. If the stop lamp lights too soon, the switch should be lowered, and vice-versa.
4  Should excessive wear be evident in the rear brake, it should be dismantled and examined, referring to Section 4 of this Chapter.

## 19 Cush drive assembly: examination and renovation

1  The cush drive assembly is contained within the left-hand side of the rear wheel hub. It comprises a set of synthetic rubber buffers, housed within a series of vanes cast in the hub shell. A plate attached to the centre of the rear wheel sprocket has four cast-in dogs which engage with slots in these rubbers, when the wheel is replaced in the frame. The drive to the rear wheel is transmitted via these rubbers, which cushion any surges or roughness in the drive which would otherwise convey the impression of harshness.
2  Examine the rubbers periodically for signs of damage or general deterioration. Renew and fit the rubbers as a set if there is any doubt about their condition; there is no difficulty in removing or replacing them as they are not under compression when the drive plate is attached.

## 20 Rear wheel sprocket: removal, examination and replacement

1  The rear wheel sprocket assembly can be removed as a separate unit after the rear wheel has been separated and detached from the frame as described in Section 17 of this Chapter. Alternatively, it can be removed still attached to the rear wheel if the final drive chain is detached and the procedure described in Chapter 4, Section 11 is followed.
2  Check the condition of the sprocket teeth. If they are hooked, chipped or badly worn, the sprocket must be renewed. It is secured to the cush drive plate by four bolts and lock washers.
3  It is considered bad practice to renew one sprocket on its own. The final drive sprockets should always be renewed as a pair and a new chain fitted, otherwise rapid wear will necessitate even earlier renewal on the next occasion.
4  An additional bearing is located within the cush drive plate, which supports the sprocket shaft into which the rear wheel spindle fits. In common with the wheel bearings, this bearing is a journal ball and when wear occurs, the sprocket will give the appearance of being loose on its mounting bolts. The bearing is a tight push fit on the sprocket shaft and is preceded by an oil seal that excludes road grit and water. A circlip retains the assembly within the cush drive plate.
5  Remove the oil seal and bearing and wash out the latter to remove all traces of the old grease. If the bearing has any play or runs roughly, it must be renewed.
6  If the bearing has not been renewed it should be repacked with grease and pushed back on the sprocket shaft, followed by the oil seal. Replace the rear wheel assembly by reversing whichever method was adopted for its removal.

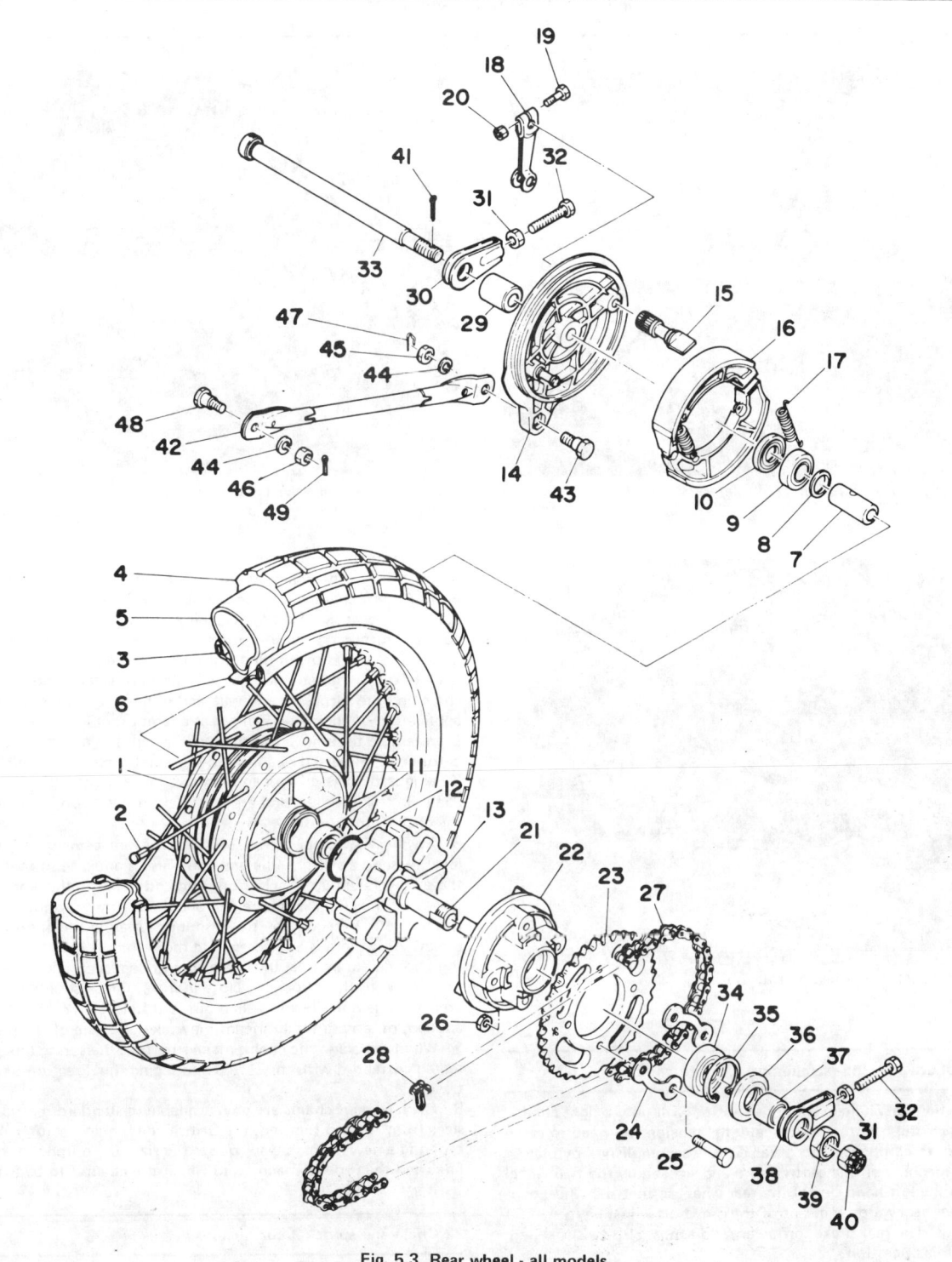

**Fig. 5.3. Rear wheel - all models**

| | | | |
|---|---|---|---|
| 1  Hub | 13  Cush drive rubber - 4 off | 26  Nut - 2 off | 38  Adjuster |
| 2  Spoke set | 14  Brake plate | 27  Rear chain | 39  Outer spindle nut |
| 3  Wheel rim | 15  Fulcrum | 28  Joining link | 40  Wheel spindle nut |
| 4  Tyre | 16  Brake shoe - 2 off | 29  Spacer | 41  Split pin |
| 5  Inner tube | 17  Spring - 2 off | 30  Adjuster | 42  Torque arm |
| 6  Rim tape | 18  Actuating lever | 31  Locknut - 2 off | 43  Special bolt |
| 7  Bearing spacer | 19  Pinch bolt | 32  Bolt - 2 off | 44  Spring washer - 2 off |
| 8  Spacer | 20  Nut | 33  Wheel spindle | 45  Nut |
| 9  Bearing | 21  Outer spindle | 34  Bearing | 46  Nut |
| 10  Oil seal | 22  Cush drive boss | 35  Circlip | 47  R pin |
| 11  Seal | 23  Rear wheel sprocket | 36  Seal | 48  Bolt |
| 12  O ring | 24  Lock washer - 2 off | 37  Spacer | 49  Split pin |
| | 25  Bolt - 2 off | | |

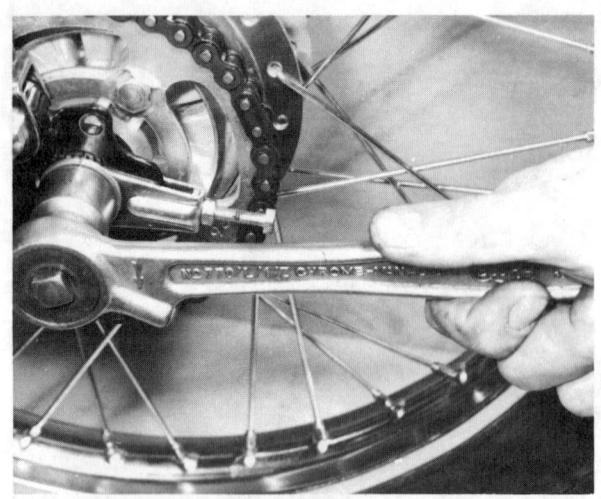

Chapter 5: Wheels, brakes and tyres

19.1a Large nut retains outer spindle

19.1b Fit new cushdrive rubbers, if worn

20.2 Sprocket is retained by four bolts

## 21 Final drive chain: examination and lubrication

1 The final drive chain is fully exposed, with only a light chainguard over the top run. Periodically the tension will need to be adjusted, to compensate for wear. This is accomplished by placing the machine on the centre stand and slackening the two wheel nuts on the left-hand side of the rear wheel so that the wheel can be drawn backward by means of the drawbolt adjusters in the fork ends. The rear brake torque arm bolt must also be slackened during this operation.

2 The chain is in correct tension if there is approximately 20 mm (¾ in) slack in the middle of the lower run. Always check when the chain is at its tightest point as a chain rarely wears evenly during service.

3 Always adjust the drawbolts an equal amount in order to preserve wheel alignment. The forks ends are clearly marked with a series of vertical lines above the adjusters, to provide a simple, visual check. If desired, wheel alignment can be checked by running a straight plank of wood parallel to the machine, so that it touches the side of the rear tyre. If wheel alignment is correct, the plank will be equidistant from each side of the front wheel tyre, when tested on both sides of the rear wheel. It will not touch the front wheel tyre because this tyre is of smaller cross section.

See accompanying diagram.

4 Do not run the chain overtight to compensate for uneven wear. A tight chain will place undue stress on the gearbox and rear wheel bearings, leading to their early failure. It will also absorb a surprising amount of power.

5 After a period of running, the chain will require lubrication. Lack of oil will greatly accelerate the rate of wear of both the chain and the sprockets and will lead to harsh transmission. The application of engine oil will act as a temporary expedient, but it is preferable to remove the chain and clean it in a paraffin bath before it is immersed in a molten lubricant such as 'Linklife' or 'Chainguard'. These lubricants achieve better penetration of the chain links and rollers and are less likely to be thrown off when the chain is in motion.

6 To check whether the chain is due for replacement, lay it lengthwise in a straight line and compress it endwise so that all the play is taken up. Anchor one end and measure the length. Now pull the chain with one end anchored firmly, so that the chain is fully extended by the amount of play in the opposite direction. If there is a difference of more than ¼ inch per foot in the two measurements, the chain should be replaced in conjunction with the sprockets. Note that this check should be made **after** the chain has been washed out, but **before** any lubricant is applied, otherwise the lubricant may take up some of the play.

7 When replacing the chain, make sure that the spring link is seated correctly, with the closed end facing the direction of travel.

8 Replacement chains are now available in standard metric sizes from Renold Limited, the British chain manufacturer. When ordering a new chain, always quote the size, the number of chain links and the type of machine to which the chain is to be fitted.

## 22 Chain and sprocket conversions

1 It is now possible to obtain a British-made chain and sprocket conversion kit of Renold manufacture which permits non-metric components to be substituted. The main advantage is the use of a heavier weight of chain and wider sprockets, reducing the rate of wear by a significant amount. It follows that the kit must be purchased as a whole, since metric and non-metric parts cannot be used in conjunction with one another.

2 Despite widespread belief to the contrary, there is rarely any advantage to be gained from varying sprocket sizes, from those specified by the manufacturer. A large gearbox sprocket by no means guarantees a higher maximum speed; usually it converts top gear into little more than an overdrive and may even lead to a situation where the machine is faster in the next lower gear.

## Chapter 5: Wheels, brakes and tyres

21.2 Check chain for free play

21.5 Aerosol lubricant will prolong chain life

21.7 Closed end of link must face direction of travel

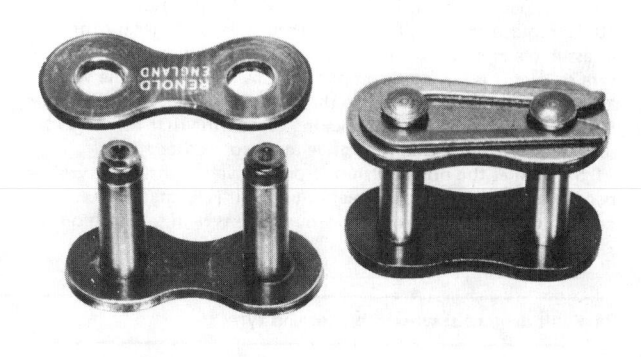

22.1 Equivalent British made chain can be obtained

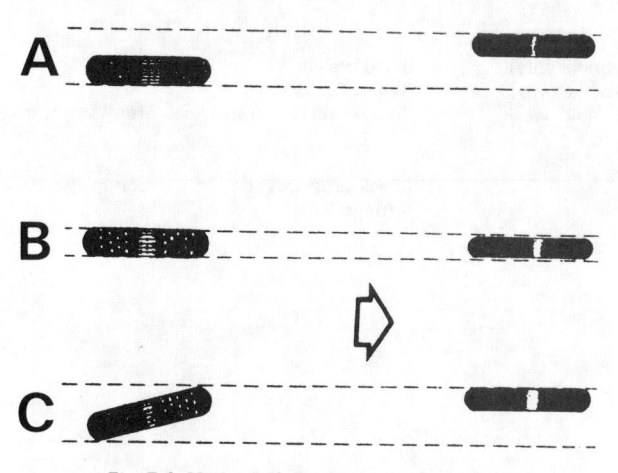

Fig. 5.4. Method of checking wheel alignment

A and C    incorrect    B correct

### 23 Tyres: removal and replacement

1  At some time or other the need will arise to remove and replace the tyres, either as the result of a puncture or because a renewal is required to offset wear. To the inexperienced, tyre changing represents a formidable task yet if a few simple rules are observed and the technique learned, the whole operation is surprisingly simple.

2  To remove the tyre from the wheel, first detach the wheel from the machine by following the procedure in Section 3 or Section 17 of this Chapter, depending on whether the front or the rear wheel is involved. Deflate the tyre by removing the valve insert and when it is fully deflated, push the bead of the tyre away from the wheel rim on both sides so that the bead enters the centre well of the rim. Remove the locking cap and push the tyre valve into the tyre itself.

3  Insert a tyre lever close to the valve and lever the edge of the tyre over the outside of the wheel rim. Very little force should be necessary; if resistance is encountered it is probably due to the fact that the tyre beads have not entered the well of the wheel rim all the way round the tyre.

4 Once the tyre has been edged over the wheel rim, it is easy to work around the wheel rim so that the tyre is completely free on one side. At this stage, the inner tube can be removed.

5 Working from the other side of the wheel, ease the other edge of the tyre over the outside of the wheel rim furthest away. Continue to work around the rim until the tyre is free from the rim.

6 If a puncture has necessitated the removal of the tyre, reinflate the inner tube and immerse it in a bowl of water to trace the source of the leak. Mark its position and deflate the tube. Dry the tube and clean the area around the puncture with a petrol-soaked rag. When the surface has dried, apply rubber solution and allow this to dry before removing the backing from the patch and applying the patch to the surface.

7 It is best to use a patch of the self-vulcanising type, which will form a permanent repair. Note that it may be necessary to remove a protective covering from the top surface of the patch, after it has sealed in position. Inner tubes made from synthetic rubber may require a special type of patch and adhesive, if a satisfactory bond is to be achieved.

8 Before replacing the tyre, check the inside to make sure the agent that caused the puncture is not trapped. Check also the outside of the tyre, particularly the tread area, to make sure nothing is trapped that may cause a further puncture.

9 If the inner tube has been patched on a number of past occasions, or if there is a tear or large hole, it is preferable to discard it and fit a new one. Sudden deflation may cause an accident, particularly if it occurs with the front wheel.

10 To replace the tyre, inflate the inner tube sufficiently for it to assume a circular shape but only just. Then push it into the tyre so that it is enclosed completely. Lay the tyre on the wheel at at an angle and insert the valve through the rim tape and the hole in the wheel rim. Attach the locking cap on the first few threads, sufficient to hold the valve captive in its correct location.

11 Starting at the point furthest from the valve, push the tyre bead over the edge of the wheel rim until it is located in the central well. Continue to work around the tyre in this fashion until the whole of the one side of the tyre is on the rim. It may be necessary to use a tyre lever during the final stages.

12 Make sure there is no pull on the tyre valve and again commencing with the area furthest from the valve, ease the other bead of the tyre over the edge of the rim. Finish with the area close to the valve, pushing the valve up into the tyre until the locking cap touches the rim. This will ensure the inner tube is not trapped when the last section of the bead is edged over the rim with a tyre lever.

13 Check that the inner tube is not trapped at any point. Re-inflate the inner tube, and check that the tyre is seating correctly around the wheel rim. There should be a thin rib moulded around the wall of the tyre on both sides, which should be equidistant from the wheel rim at all points. If the tyre is unevenly located on the rim, try bouncing the wheel when the tyre is at the recommended pressure. It is probable that one of the beads has not pulled clear of the centre well.

14 Always run the tyres at the recommended pressures and never under or over-inflate. The correct pressures for solo use are given in the Specifications Section of this Chapter.

15 Tyre replacement is aided by dusting the side walls, particularly in the vicinity of the beads, with a liberal coating of french chalk. Washing-up liquid can also be used to good effect, but this has the disadvantage of causing the inner surfaces of the wheel rim to rust.

16 Never replace the inner tube and tyre without the rim tape in position. If this precaution is overlooked there is a good chance of the ends of the spoke nipples chafing the inner tube and causing a crop of punctures.

17 Never fit a tyre that has a damaged tread or side walls. Apart from the legal aspects, there is a very great risk of a blow-out which can have serious consequences on any two-wheel vehicle.

18 Tyre valves rarely give trouble, but it is always advisable to check whether the valve itself is leaking before removing the tyre. Do not forget to fit the dust cap, which forms an effective second seal.

## 24 Fault diagnosis: wheels, brakes and tyres

| Symptom | Cause | Remedy |
| --- | --- | --- |
| Handlebars oscillate at low speeds | Buckled front wheel<br>Incorrectly fitted front tyre | Remove wheel for specialist attention.<br>Check whether line around bead is equidistant from rim. |
| Forks 'hammer' at high speeds | Front wheel out of balance | Add weights until wheel will stop in any position. |
| Brakes feel spongy | Air in hydraulic line (disc brake only)<br>Fluid leak in system (disc brake only)<br>Stretched brake operating cable, weak pull-off springs | Bleed brakes.<br>Renew faulty part.<br>Renew cable and/or springs, after inspection.. |
| Tyres wear more rapidly in middle of tread | Over inflation | Check pressures and run at recommended settings. |
| Tyres wear rapidly at outer edges of tread | Under inflation | Ditto. |

# Chapter 6 Electrical system

*For information relating to 1977 on models refer to Chapter 7*

## Contents

| | |
|---|---|
| General description ... 1 | Flashing indicator lamps ... 10 |
| Checking the electrical system: general ... 2 | Flasher unit: location and replacement ... 11 |
| Charging system: checking the output ... 3 | Speedometer and tachometer heads: replacement of bulbs ... 12 |
| Rectifier: checking for malfunctions ... 4 | |
| Battery: charging procedure ... 5 | Stop lamp switches: location and replacement ... 13 |
| Fuse: location and replacement ... 6 | Horn: location and examination ... 14 |
| Headlamp: replacing bulbs and adjusting beam height ... 7 | Wiring: layout and examination ... 15 |
| Handlebar switches: function and replacement ... 8 | Ignition and lighting switch ... 16 |
| Stop and tail lamp: replacing the bulb ... 9 | Fault diagnosis: electrical system ... 17 |

## Specifications

### Battery
- Make ... FB or GS
- Model ... 6N4A - 4D or 6N4 - 2A - 2
- Voltage ... 6 volt
- Capacity ... 4 AH

### Generator
| | RS100 | RS125 |
|---|---|---|
| Make ... | Mitsubishi | Hitachi |
| Type ... | FOOOTO 2771 | F130 - 09 |
| Voltage ... | 6 volt | 6 volt |

### Rectifier
- Make ... Stanley DE 2304 or DE 5404 half-wave silicon rectifier 6 amp

### Bulbs
- Voltage ... 6 volt
- Headlamp ... 25/25 watts
- Pilot lamp ... 4 watts*
- Stop/tail lamp ... 5.3/17 watts
- Instrument illumination lamp ... 3 watts
- Indicator warning lamp ... 3 watts
- Neutral indicator lamp ... 3 watts
- Oil level warning lamp ... 3 watts (where fitted)
- Flashing indicator lamp ... 8 watts

*On some models a 3 watt bulb may be fitted. Check bulb markings carefully prior to renewal

### Flasher relay
- Capacity ... 45 watt or 23 watt

### Horn ... 6 volt

### Fuse ... 10 amp

## Chapter 6: Electrical system

### 1 General description

1 The Yamaha RS models are equipped with a flywheel generator which feeds the 6 volt electrical system. AC current from the generator passes to a half-wave silicon rectifier, where it is converted to DC before being fed to the battery. The 6 volt 4 AH battery powers the ignition circuit and also the electrical system in general.

### 2 Checking the electrical system: general

Many of the test procedures applicable to motorcycle electrical systems require the use of test equipment of the multimeter type. Although the tests themselves are quite straightforward, there is a danger of damaging certain components if wrong connections are made. It is recommended, therefore, that no attempt be made to investigate faults in the charging system, unless the owner is reasonably experienced in this field. A qualified Yamaha Service Agent willhave in his possession the necessary diagnostic equipment to effect an economical repair.

### 3 Charging system - checking the output

1 A defect in the charging system will produce poor headlamp performance, and probably, repeatedly discharged batteries. In order to check that the battery is receiving the correct charge, the following procedure should be adopted.
2 To check the charging current, connect a multimeter set on DC amps between the rectifier and the battery. The positive probe should be attached to the red cable's terminal on the rectifier, the negative probe to the battery positive terminal. Switch on the ignition and run the engine at 8000 rpm, at which point the meter should indicate 4.5 amps or less. If the lights are now switched on, a current of $1.5 \pm 0.5$ amps should be shown for RS100 models, and $1.3 \pm 0.3$ amps in the case of the RS125. Again this figure is to be expected at 8000 rpm. These readings may vary slightly if the battery is in a discharged state, but any great divergence indicates a breakdown in the charging circuit. If this is the case, a qualified Yamaha Service Agent should be consulted. He will be able to isolate the faulty component, and supply a replacement unit.

### 4 Rectifier: checking for malfunctions

1 If the rectifier is suspected of malfunctioning it can be checked very easily using a multimeter set on the resistance scale. Connect the meter's red probe (+) to the red terminal of the rectifier, and the black probe (−) to the white rectifier terminal. For the purposes of this test, the rectifier leads should be detached. With the test probes connected as described above, a reading of 9 - 10 ohms should be obtained. Reversing the probes should indicate no continuity. If this is not the case, the rectifier can be assumed to be defective and should be renewed.

### 5 Battery: charging procedure

1 The normal charging rate for the 4 amp hour battery is 0.4 amps. A more rapid charge, not exceeding 1 amp can be given in an emergency. The higher charge rate should, if possible, be avoided since it will shorten the working life of the battery.
2 Make sure that the battery charger connections are correct, red to positive and black to negative. It is preferable to remove the battery from the machine whilst it is being charged and to remove the vent plug from each cell. When the battery is reconnected to the machine, the black lead must be connected to the negative terminal and the red lead to positive. This is most important, as the machine has a negative earth system. If the terminals are inadvertently reversed, the electrical system will be damaged permanently. The rectifier will be destroyed by a reversal of the current flow.
3 If the machine is used exclusively for short trips, it may be found that frequent recharging is necessary due to the generator output being insufficient to cope with the machine's electrical demands. This may be particularly noticeable if the machine is used only for travelling to work, when, during the winter months the lights and especially the stop light, will be used almost continuously. Provided that the battery and generator are known to be in good order, it is permissible to change the generator output wiring as follows to obtain a higher charge rate:
4 Disconnect the green output lead at the bullet connector. The generator side should be isolated with insulating tape or similar, and the switch side should now be connected to the green/red output lead. The resulting higher charge rate will help to offset the heavy drain on the electrical system, but a watchful eye must be kept on the battery for signs of overcharging. If the electrical demand falls again, ie, during the summer months, the original connections must be restored to prevent damage to the battery.

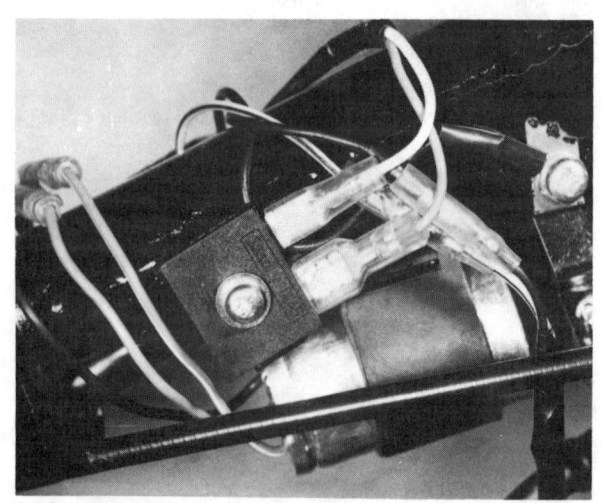

4.1 Rectifier is bolted to top frame tube

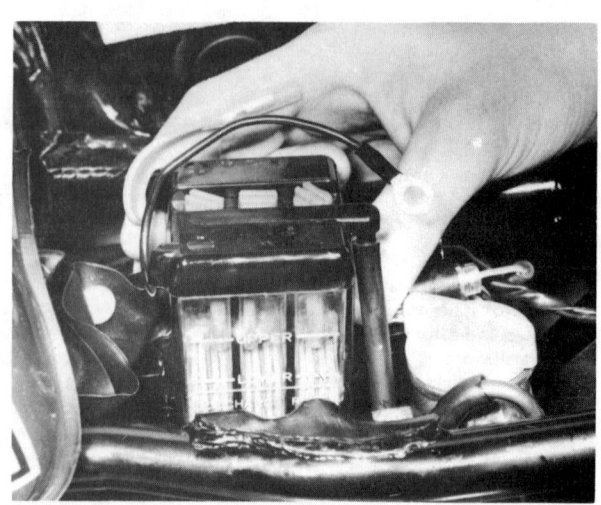

5.2 Battery is located beneath dualseat

# Chapter 6: Electrical system

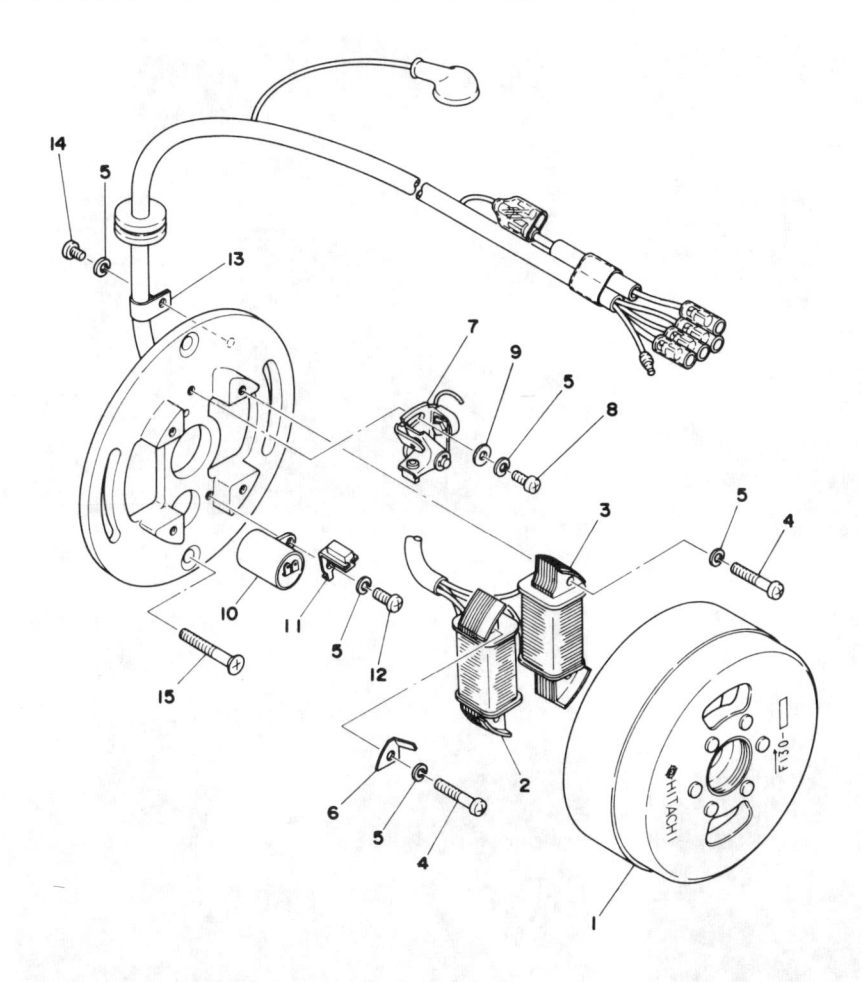

**Fig. 6.1. Flywheel generator - component parts**

1 Flywheel assembly
2 Generator coil
3 Generator coil
4 Screw - 4 off
5 Spring washer - 7 off
6 Plate
7 Contact breaker assembly
8 Screw
9 Washer
10 Condenser
11 Lubricator wick
12 Screw
13 Clamp
14 Screw
15 Countersunk screw - 2 off

## 6 Fuse: location and replacement

1 The main fuse is fitted in line in the battery positive lead. It is located in a white plastic fuseholder fitted beneath the fuel tank and dualseat and is rated at 10 amps.

2 Before replacing a fuse that has blown, check that no obvious short circuit has occurred, otherwise the replacement fuse will blow immediately it is inserted. It is always wise to check the electrical circuit thoroughly, to trace the fault and eliminate it.

3 When a fuse blows while the machine is running and no spare is available, a 'get you home' remedy is to remove the blown fuse and wrap it in silver paper before replacing it in the fuseholder. The silver paper will restore the electrical continuity by bridging the broken fuse wire. This expedient should **never** be used if there is evidence of a short circuit or other major electrical fault, otherwise more serious damage will be caused. Replace the 'doctored' fuse at the earliest possible opportunity, to restore full circuit protection.

## 7 Headlamp: replacing bulbs and adjusting beam height

1 To remove the headlamp rim, detach the small screw on the right-hand underside of the headlamp shell. The rim can then be prised off, complete with the reflector unit.

2 The main bulb is a twin filament type, to give a dipped beam facility. The bulb holder is attached to the back of the reflector by a rubber sleeve, which fits around a flange in the reflector and the flange of the bulb holder. An indentation in the bulb holder orifice and a projection on the bulb holder end ensures the bulb is always replaced in the same position so that the focus is unaltered.

3 It is not necessary to re-focus the headlamp when a new bulb is fitted. Apart from the just-mentioned method of location, the bulbs used are of the pre-focus type, built to a precise specification. To release the bulb holder, twist and lift away.

4 The pilot lamp bulb, like the bulbholder has a bayonet fitting, It is protected by a rubber sleeve. Remove the bulb holder first, then the bulb.

5 The main headlamp bulb is rated at 25/25W, 6 volts and the pilot lamp bulb at 4W, 6 volts. Variations in the wattage may occur according to the country or state for which the machine is supplied.

6 Beam alignment is adjusted by tilting the headlamp after the two retaining bolts have been slackened and then retightening them after the correct beam height is obtained, without moving the setting.

7 To obtain the correct beam height, place the machine on level ground facing a wall 25 feet distant, with the rider seated normally. The height of the beam centre should be equal to the height of the centre of the headlamp from the ground, when the dip switch is in the main beam position. Furthermore, the concentrated area of light should be centrally disposed. Adjustments in either direction are made by rearranging the angle of the headlamp, as described in the preceding paragraph. Note that a different beam setting will be needed when a pillion passenger is carried. If a pillion passenger is carried regularly, the passenger should be seated in addition to the rider when the beam setting adjustment is made.

8 The above instructions for beam setting relate to the requirements of the United Kingdom's transport lighting regulations. Other settings may be required in countries other than the UK.

Chapter 6: Electrical system

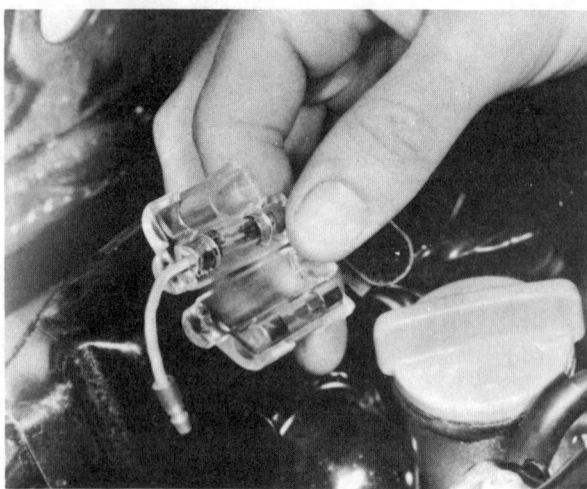

6.1 Fuseholder is adjacent to battery (Note spare fuse)

6.7a Remove screw to release ...

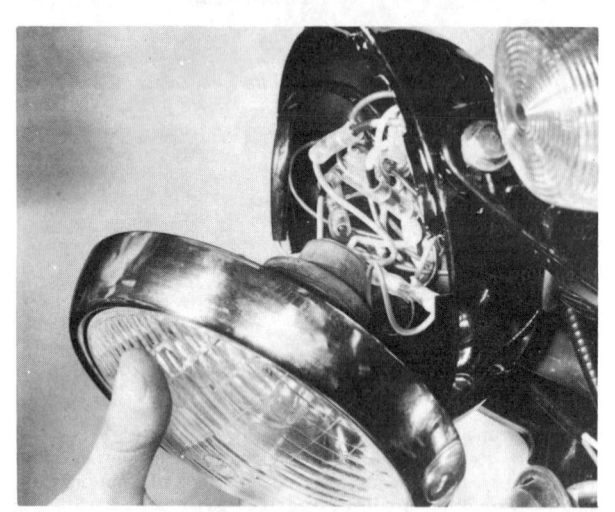

7.1b ...headlamp unit from shell

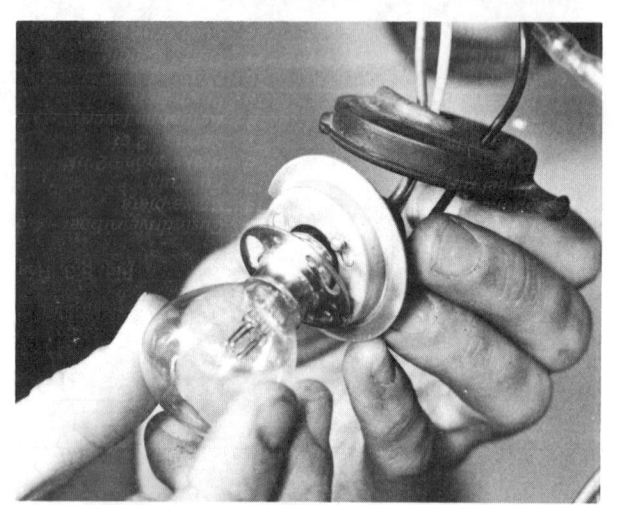

7.2 Bulb is retained by rubber shroud

### 8 Handlebar switches: function and replacement

1 The dipswitch forms part of the left-hand dummy twist grip which contains the horn button and the indicator lamp switch. The right-hand twist grip assembly incorporates the lighting master switch and a three position ignition positive cut-out switch.

2 In the event of failure of any of these switches, the switch assembly must be replaced as a complete unit since it is not practicable to effect a permanent repair.

Before condemning the unit, however, check that the malfunction is not due to dirty or burnt contacts. It may be possible to restore these with a switch cleaning spray or by burnishing them with fine wet or dry paper.

### 9 Stop and tail lamp: replacing the bulb

1 The tail lamp is fitted with a twin filament bulb of 6 volt, 5.3/17W rating, to illuminate the rear number plate and rear of the machine, and to give visual warning when the rear brake is applied. To gain access to the bulb remove the plastic lens cover, which is retained by two long screws. Check that the gasket between the lens cover and the main body of the lamp is in good condition.

2 The bulb has a bayonet fitting and has staggered pins to prevent the bulb contacts from being reversed.

3 If the tail lamp bulb keeps blowing, suspect either vibration of the rear mudguard or more probably an intermittent earth connection.

### 10 Flashing indicator lamps

1 The forward facing indicator lamps are connected to 'stalks' which are attached to the ends of the top fork yoke. The hollow stalks carry the leads to the lens unit. The rear facing lamps are mounted on similar shorter stalks, at a point immediately to the rear of the dualseat.

2 In each case, access to the bulb is gained by removing the plastic lens cover, which is retained by two screws. Bayonet fitting bulbs of the single filament type are used, each with a 6 volt 8 watt rating

## Chapter 6: Electrical system 97

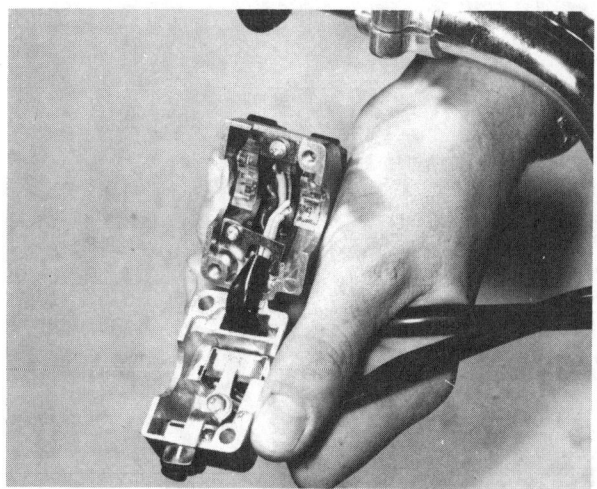

8.2 Switch contacts can be cleaned with aerosol spray

9.1 Stop/tail bulb has offset bayonet fitting

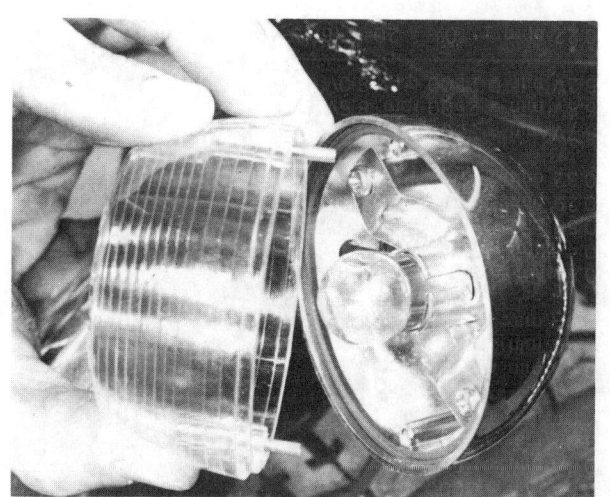

10.2a Remove lens unit to gain ...

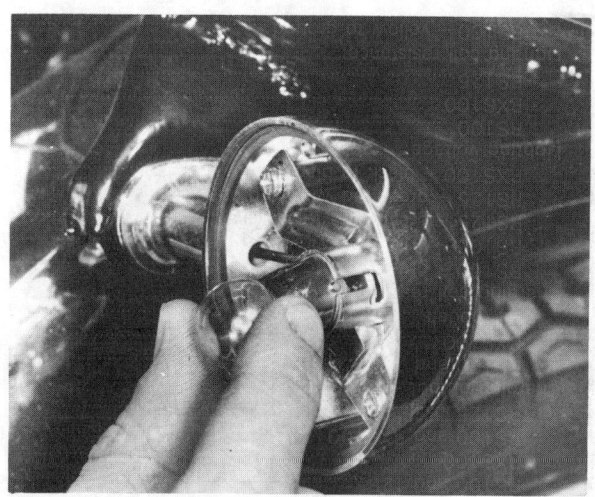

10.2b ... access to bulb

### 11 Flasher unit: location and replacement

1  The flasher relay unit is located under the dualseat, being rubber-mounted to the frame top tube.
2  If the flasher unit is functioning correctly, a series of audible clicks will be heard when the indicator lamps are in action. If the unit malfunctions and all the bulbs are in working order, the usual symptom is one initial flash before the unit goes dead; it will be necessary to replace the unit complete if the fault cannot be attributed to any other cause.
3  Take great care when handling the unit because it is easily damaged if dropped.

### 12 Speedometer and tachometer heads: replacement of bulbs

1  The speedometer head (and tachometer where fitted) houses an illumination bulb rated at 6 volts 3W, and the neutral and indicator warning lamps, these also being rated at 6 volts 3W
2  Access to the lamps is obtained by removing the relevant drive cable and the two pins retaining the instrument to its mounting. The instrument can then be lifted away and the rubber bulb-holders pulled out.

### 13 Stop lamp switches: location and replacement

1  Two stop lamp switches are fitted to the machine, which work independently of one another, depending on which brake is operated.
2  The front brake switch is fitted to the handlebar lever stock and is a mechanical push-off type, being operated when the lever is moved. The switch is a push fit in the housing boss, and is detached by depressing a small pin in the underside with a piece of wire or a small screwdriver.
3  The rear brake switch is mounted on the frame on the right-hand side, above the rear brake pedal. It can be adjusted by means of a locknut, and should be set so that the light comes on as soon as the pedal is depressed. This is especially important when the rear brake has been readjusted.

Chapter 6: Electrical system

11.1 Flasher relay is rubber-mounted beneath tank

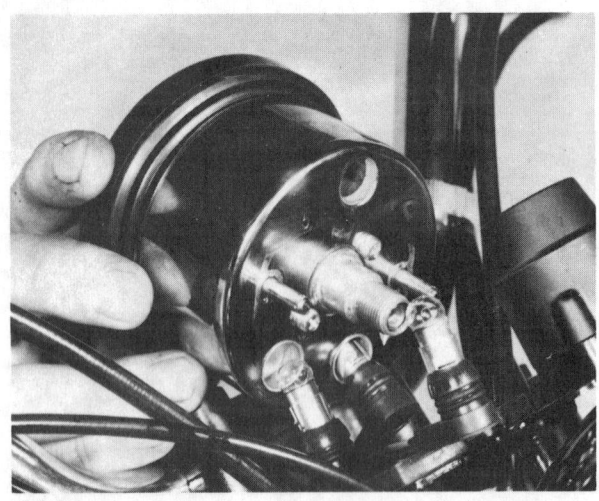

12.2 Remove instrument head to gain access to bulbs

13.2 Front brake switch is push-fit in lever

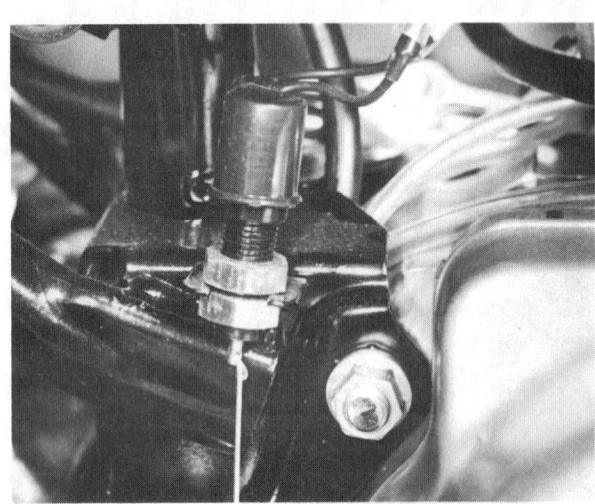

13.3 Rear brake switch can be adjusted for height

## 14 Horn: location and examination

1  The horn is located between the fork legs, bolted to the bottom yoke. It has no external means of adjustment. If it malfunctions, it must be renewed; it is a statutory requirement that the machine must be fitted with a horn in working order.

## 15 Wiring: layout and examination

1  The wiring harness is colour-coded and will correspond with the accompanying wiring diagram. Where socket connectors are used, they are designed so that reconnection can be made in the correct position only.
2  Visual inspection will show whether there are any breaks or frayed outer coverings which will give rise to short circuits. Another source of trouble may be the snap connectors and sockets, where the connector has not been pushed fully home in the outer housing.

14.1 Horn is mounted between fork legs

## Chapter 6: Electrical system

3 Intermittent short circuits can often be traced to a chafed wire that passes through or is close to a metal component such as a frame member. Avoid tight bends in the lead or situations where a lead can become trapped between casings.

### 16 Ignition and lighting switch

1 The ignition and lighting switch is combined in one unit, bolted to the top fork yoke. It is operated by a key, which cannot be removed when the ignition is switched on.

2 The number stamped on the key will match the number of the steering head lock and that of the lock in the petrol filler cap. A replacement key can be obtained if the number is quoted; if either of the locks or the ignition switch is changed, additional keys will be required.

3 It is not practicable to repair the ignition switch if it malfunctions. It should be renewed with a new switch and key to suit.

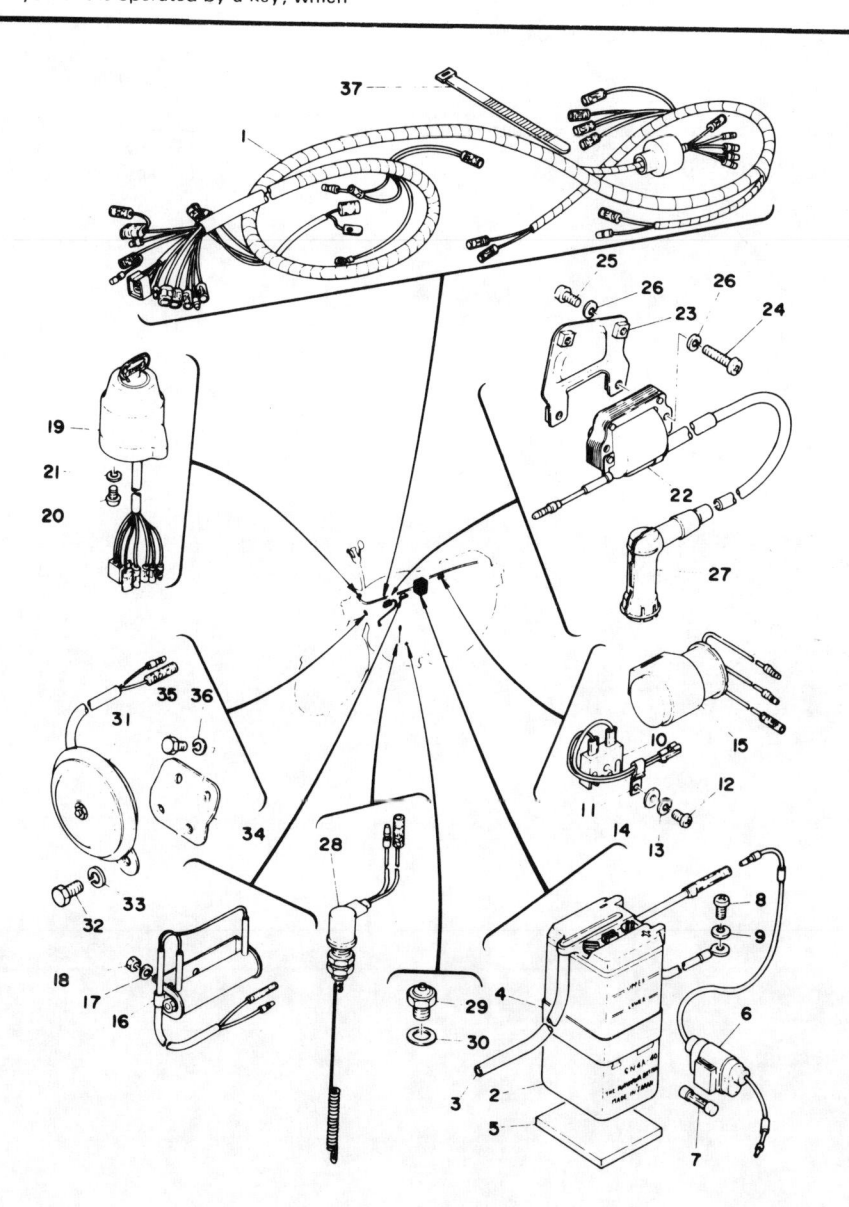

Fig. 6.2. Electrical system - component location

| | | | | | | | |
|---|---|---|---|---|---|---|---|
| 1 | Wiring harness | 10 | Rectifier | 19 | Ignition/lighting switch | 28 | Rear brake light switch |
| 2 | Battery | 11 | Clamp | 20 | Screw | 29 | Neutral indicator switch |
| 3 | Vent pipe | 12 | Screw | 21 | Spring washer - 2 off | 30 | Fibre washer |
| 4 | Strap | 13 | Spring washer | 22 | Ignition coil | 31 | Horn |
| 5 | Mounting pad | 14 | Plain washer | 23 | Mounting bracket | 32 | Bolt - 2 off |
| 6 | Fuse holder | 15 | Flasher relay | 24 | Screw - 2 off | 33 | Spring washer - 2 off |
| 7 | Fuse (6v 10 amp) - 2 off | 16 | Resistor | 25 | Screw - 2 off | 34 | Bracket |
| 8 | Screw | 17 | Spring washer | 26 | Spring washer - 4 off | 35 | Bolt - 2 off |
| 9 | Washer | 18 | Nut | 27 | Spark plug cap | 36 | Spring washer - 2 off |
| | | | | | | 37 | Cable tie |

## 17 Fault diagnosis: electrical system

| Symptom | Cause | Remedy |
| --- | --- | --- |
| Complete electrical failure | Blown fuse | Check wiring and electrical components for short circuit before fitting new 10 amp fuse. Check battery connections, also whether connections show signs of corrosion. |
| Dim lights, horn inoperative | Discharged battery | Recharge battery with battery charger and check whether alternator is giving correct output (electrical specialist). |
| Constantly 'blowing' bulbs | Vibration, poor earth connection | Check whether bulb holders are secured correctly. Check earth return or connections to frame. |

101

Right-hand view of the Yamaha RXS100

Engine/gearbox unit of the Yamaha RXS100

# Chapter 7 The 1977 on models

**Contents**

| | |
|---|---|
| Introduction | 1 |
| Cylinder head and barrel: removal and refitting – RXS100 | 2 |
| Crankcase covers: removal | 3 |
| Tachometer drive: cable removal | 4 |
| Kickstart shaft: removal and refitting | 5 |
| Gearchange linkage: removal and refitting | 6 |
| Crankshaft assembly: modification | 7 |
| Gearbox layshaft: modification – RXS100 | 8 |
| Gear selector forks: modification – RXS100 | 9 |
| Petrol tap: removal and refitting – RXS100 | 10 |
| Carburettor: modifications – RS100 and 125 | 11 |
| Carburettor: modifications – RXS100 | 12 |
| Air cleaner: removal, cleaning and replacement | 13 |
| YEIS unit: removal and refitting – RXS100 | 14 |
| Exhaust system: removal and refitting – RXS100 | 15 |
| Oil pump: drive pinion removal | 16 |
| Oil level gauge and warning lamp | 17 |
| Electronic ignition system – RXS100 | 18 |
| Steering head: modifications | 19 |
| Front fork legs: dismantling and reassembly | 20 |
| Swinging arm: pivot bolt – RS100 1981 and 1982 | 21 |
| Swinging arm: bush removal – RXS100 | 22 |
| Front brake caliper: pad renewal – RS125 1981 on | 23 |
| Front brake caliper: dismantling and reassembly – RS125 1981 on | 24 |
| Rear wheel: modifications – RXS100 | 25 |
| Headlamp: bulb renewal – RXS100 | 26 |
| Speedometer and tachometer heads: bulb renewal | 27 |

**Specifications**

Except where entered below, specifications for the models covered in this Chapter are the same as given for the earlier models at the beginning of each Chapter. Note the following abbreviations:
N/Av     Not available
N/App    Not applicable

**Specifications relating to Chapter 1**

**Engine – RXS100**
Capacity ............................................................ 98cc
Bore .................................................................. 50 mm (1.97 in)
Stroke ............................................................... 50 mm (1.97 in)
Compression ratio ............................................. N/Av

# Chapter 7 The 1977 on models

**Piston ring end gap (installed)**
  RS100 and 125 ............... 0.3 – 0.5 mm (0.012 – 0.020 in)
  RXS100 ............... 0.15 – 0.35 mm (0.006 – 0.014 in)

**Clutch**
  Type ............... Wet, multi plate
  No. of plain plates ............... 4
  No. of friction plates ............... 5
  No. of anti-drag rings – RS100 and 125 models only ............... 5
  No. of springs ............... 4
  Spring free length/wear limit:
    RS100 ............... 31.5 mm (1.24 in)/30.5 mm (1.20 in)
    RXS100 ............... 34.5 mm (1.36 in)/33.2 mm (1.31 in)
    RS125 ............... 34.0 mm (1.34 in)/33.0 mm (1.30 in)
  Friction plate thickness ............... 3.0 mm (0.120 in)
  Wear limit ............... 2.7 mm (0.106 in)

**Gearbox – All models**
  1st gear ............... 2.833 : 1
  2nd gear ............... 1.706 : 1
  3rd gear ............... 1.250 : 1
  4th gear ............... 1.091 : 1
  5th gear ............... 0.957 : 1

**Secondary reduction ratio**
  RS and RXS100 ............... 2.467 : 1
  RS125 ............... 2.400 : 1

## Specifications relating to Chapter 2

**Carburettor**

|  | RS100 | RXS100 | RS125 |
|---|---|---|---|
| Make | Mikuni | Mikuni | Mikuni |
| Type | VM20 SH | VM22 SS | VM24 SH |
| Main jet | 125 | 120 | 125 |
| Air jet | 2.5 | 1 | 0.5 |
| Jet needle | 4J21 | 402 | 512 |
| Clip position | 2 | 2 | 2 |
| Needle jet | P-2 | P-0 | E-4 |
| Pilot jet | 20 | 17.5 | 30 |
| Air screw turns out | 1¾ | 2¼ | 1¾ |
| Starting jet | 30 | 30 | 40 |
| Float height | 21 ± 1 mm | 21 ± 1 mm | 21 ± 1 mm |
| Idle speed | 1300 ± 50 rpm | 1400 ± 50 rpm | 1200 ± 50 rpm |

**Reed valve – RXS100**
  Stopper plate height ............... 9 mm (0.354 in)
  Maximum bend limit ............... 0.3 mm (0.012 in)

**Fuel tank capacity – RXS100** ............... 9 litres (1.97 Imp gall)

**Transmission oil – RXS100**
  Capacity – at oil change ............... 650cc (1.14 Imp pt)
  Capacity – after overhaul ............... 700cc (1.23 Imp pt)
  Grade ............... SAE 10W/30 SE

## Specifications relating to Chapter 3

**Generator**

|  | RS100 | RXS100 | RS125 |
|---|---|---|---|
| Make | Mitsubishi | Yamaha | Hitachi |
| Model | F1T250 | 31J | F143-01 |
| Output | 6 volts | 6 volts | 6 volts |
| Source coil resistance | 2.1 ohm ± 10% at 20°C (68°F) | 240 ohm ± 10% at 20°C (68°F) | 2.7 ohm ± 10% at 20°C (68°F) |
| Pulser coil |  | 20 ohm ± 10% at 20°C (68°F) |  |

**Ignition timing**

| RS100 and 125 | RXS100 |
|---|---|
| 1.8± 0.15 mm BTDC | 20° ± 1.5° BTDC @ 5000 rpm |

## Ignition coil
| | | |
|---|---|---|
| Make | Mitsubishi | Yamaha |
| Model | F6T411 | 2T4 |
| Primary resistance | 1.0 ohm ± 10% @ 20°C (68°F) | 1.6 ohm ± 10% @ 20°C (68°F) |
| Secondary resistance | 5.9 Kohm ± 20% @ 20°C (68°F) | 6.6 Kohm ± 20% @ 20°C (68°F) |

## Spark plug
| | |
|---|---|
| Type – RXS100 | NGK BR8HS |
| Gap – all models | 0.6 – 0.7 mm (0.024 – 0.028 in) |

## Specifications relating to Chapter 4

### Front forks
| | |
|---|---|
| Oil capacity: | |
| RS100 and 125 | 125 – 135 cc (4.4 ± 0.17 fl oz) |
| RXS100 | 181 cc (6.37 fl oz) |
| Oil grade | SAE 10W/30 SE motor oil |
| Spring free length: | |
| RS100 and 125 | 295 mm (11.6 in) |
| RXS100 | 306.5 mm (12.1 in) |

Swinging arm free play ..... 1 mm (0.040 in)

## Specifications relating to Chapter 5

### Tyre pressures
| | Front | Rear |
|---|---|---|
| RS100 and 125 – normal riding | 21 psi (1.5 kg/cm$^2$) | 28 psi (2.0 kg/cm$^2$) |
| RXS 100: | | |
| 0 – 90 kg (198 lb) load | 22 psi (1.5 kg/cm$^2$) | 28 psi (2.0 kg/cm$^2$) |
| 90 kg (198 lb) – maximum load | 22 psi (1.5 kg/cm$^2$) | 32 psi (2.3 kg/cm$^2$) |

### Brakes
| | |
|---|---|
| Drum ID service limit: | |
| RS100 – front and rear | 112 mm (4.410 in) |
| RXS100 – front and rear | 111 mm (4.370 in) |
| RS125 – rear | 132 mm (5.197 in) |
| Shoe lining thickness material | 4 mm (0.157 in) |
| Service limit | 2 mm (0.079 in) |

## Specifications relating to Chapter 6

### Generator
| | |
|---|---|
| Charging output – lights off: | |
| RS100 and 125 | 0.6A min @ 3000 rpm |
| RXS100 | 0.7A min @ 2500 rpm |
| RS100 and RXS100 | 3.0A max @ 8000 rpm |
| RS125 | 4.0A max @ 8000 rpm |
| Charging output – lights on: | |
| RS100 and 125 | 0.6A min @ 3000 rpm |
| RXS100 | 0.4A min @ 2500 rpm |
| All models | 3.0A max @ 8000 rpm |
| Charging coil resistance: | |
| RS100 – green wire to ground | 0.8 ohm ± 10% @ 20°C (68°F) |
| RS100 – green/red wire to ground | 0.4 ohm ± 10% @ 20°C (68°F) |
| RS125 – green wire to ground | 0.56 ohm ± 10% @ 20°C (68°F) |
| RS125 – green/red wire to ground | 0.29 ohm ± 10% @ 20°C (68°F) |
| RXS100 – black to black/white wire | 0.32 ohm ± 10% @ 20°C (68°F) |
| Lighting voltage: | |
| RS100 | 5.7V min @ 3000 rpm, 8.0V max @ 8000 rpm |
| RXS100 | 5.8V min @ 2500 rpm, 8.5V max @ 8000 rpm |
| RS125 | 5.2V min @ 2500 rpm, 8.0V max @ 8000 rpm |
| Lighting coil resistance: | |
| RS100 – yellow to ground | 0.22 ohm ± 10% @ 20°C (68°F) |
| RXS100 – black to yellow | 0.24 ohm ± 10% @ 20°C (68°F) |
| RS125 – yellow to ground | 0.18 ohm ± 10% @ 20°C (68°F) |

## Chapter 7 The 1977 on models

**Rectifier**
- Make .................................................. Toshiba S5108
- Capacity .............................................. 4A

**Voltage regulator**
- Type .................................................. Mitsubishi TS6HRY-LB
- No load regulated voltage ................. 7.2 – 7.6 volts

**Bulbs**
- Headlamp ........................................... 6V, 35/35W
- Pilot lamp:
  - RS and RXS100 ............................... 6V, 3W
  - RS125 ............................................. 6V, 4W
- Stop/tail lamp ..................................... 6V, 21/5W
- Flashing indicator lamp:
  - 1977 – 80 RS100 and 125 ................. 6V, 10W
  - 1981 on RS100 and 125, 1983 – 86 RXS100 ... 6V, 15W
  - 1987 on RXS100 ............................... 6V, 21W
- Instrument illuminating lights, neutral light, high beam light, oil level light and flashing indicator warning lights ........ 6V, 3W

### Recommended torque settings

| | RXS100 | RS100 | RS125 |
|---|---|---|---|
| Cylinder head | 2.5 kgf m (18 lbf ft) | 2.5 kgf m (18 lbf ft) | 2.5 kgf m (18 lbf ft) |
| Generator flywheel nut | 7.0 kgf m (50 lbf ft) | 5.0 kgf m (36 lbf ft) | 5.0 kgf m (36 lbf ft) |
| Clutch centre nut | 4.5 kgf m (32 lbf ft) | 4.5 kgf m (32 lbf ft) | 4.5 kgf m (32 lbf ft) |
| Gearbox sprocket nut | 5.5 kgf m (40 lbf ft) | 5.5 kgf m (40 lbf ft) | 5.5 kgf m (40 lbf ft) |
| Crankcase screws | 0.7 kgf m (5.1 lbf ft) | 1.0 kgf m (7.5 lbf ft) | 1.0 kgf m (7.5 lbf ft) |
| Crankcase cover screws | 0.9 kgf m (6.5 lbf ft) | N/Av | N/Av |
| Crankshaft pinion nut | 5.0 kgf m (36 lbf ft) | 6.0 kgf m (43.5 lbf ft) | 6.0 kgf m (43.5 lbf ft) |
| Transmission oil drain plug | 2.0 kgf m (14 lbf ft) | N/Av | N/Av |
| Front wheel spindle nut | 4.2 kgf m (30 lbf ft) | 11 kgf m (80 lbf ft) – disc brake model 4.5 kgf m (32 lbf ft) – drum brake model | 11 kgf m (80 lbf ft) |
| Front wheel spindle clamp | N/App | 2.0 kgf m (14 lbf ft) | 2.0 kgf m (14 lbf ft) |
| Fork yoke pinch bolts: | | | |
| Upper | 1.5 kgf m (11 lbf ft) | 1.0 kgf m (7.5 lbf ft) | 1.0 kgf m (7.5 lbf ft) |
| Lower | 3.2 kgf m (23 lbf ft) | 3.5 kgf m (25 lbf ft) | 3.5 kgf m (25 lbf ft) |
| Engine mounting bolts: | | | |
| Front | 4.0 kgf m (29 lbf ft) | 2.5 kgf m (18 lbf ft) | 2.5 kgf m (18 lbf ft) |
| Rear upper | 5.8 kgf m (44 lbf ft) | 4.0 kgf m (30 lbf ft) | 4.0 kgf m (30 lbf ft) |
| Rear lower | 5.8 kgf m (44 lbf ft) | 5.0 kgf m (36 lbf ft) | 5.0 kgf m (36 lbf ft) |
| Swinging arm pivot shaft | 5.8 kgf m (44 lbf ft) | 4.5 kgf m (33 lbf ft) | 4.5 kgf m (33 lbf ft) |
| Rear wheel spindle nut | 4.2 kgf m (30 lbf ft) | 4.5 kgf m (33 lbf ft) | 8.5 kgf m (61 lbf ft) |
| Rear wheel sprocket bolts | 2.6 kgf m (18 lbf ft) | N/Av | N/Av |
| Rear wheel outer spindle nut | 8.5 kgf m (61 lbf ft) | N/Av | N/Av |
| Rear suspension unit | 4.0 kgf m (29 lbf ft) | 3.0 kgf m (22 lbf ft) | 3.0 kgf m (22 lbf ft) |

## Chapter 7 The 1977 on models

### 1 Introduction

This Chapter covers the 1977 on RS100 and 125 models and the RXS100 model, introduced during 1983. All modifications to these models are detailed in the following sections where the working procedure differs from that given in Chapters 1 to 6, for the 1974 to 1976 models.

The RS100 and 125 models are virtually unchanged apart from modified front forks, and on later 125 models a revised front brake caliper, with detail changes only to the gearchange, carburettor and oil pump. In April 1983 the RXS100 was introduced, and eventually replaced the RS100 model. Although basically similar to its predecessor, the RXS100 featured electronic ignition, a modified swinging arm and many revised engine and fuel system components. Apart from slight modification during 1987, when the prop stand was omitted and matt black indicators fitted, this model has remained virtually unchanged since its introduction.

When working on a 1977 or later model, refer first to this Chapter to note any modifications. If no mention is made, the task will be substantially the same as given in the previous six Chapters.

### 2 Cylinder head and barrel: removal and refitting – RXS 100

1   Before removing the cylinder head it is first necessary to detach the exhaust system front pipe and carburettor. Remove the two nuts at the exhaust port flange and unscrew the large gland nut at the pipe and silencer connection. The pipe can now be detached. It is not necessary to take off the carburettor, only to disconnect the hose to the cylinder. Slacken the hose clamp and peel the hose away from the reed valve body. Squeeze the ears of the YEIS hose clip together and pull the hose off the reed valve stub union.
2   Pull the suppressor cap off the spark plug. Slacken and remove the four cylinder head flange nuts in a diagonal sequence. Lift off the head and gasket.
3   Remove the four flange nuts from the cylinder barrel base and lift the barrel off its crankcase studs, supporting the piston as it emerges from the bore. Place a clean rag in the crankcase mouth before the lower edge of the barrel frees the rings. This will preclude any small particles of broken ring falling into the crankcase.
4   Examination and renovation should be conducted as described in Chapter 1, Sections 18 to 20.
5   Before reassembly, clean all mating surfaces and have ready a new base and head gasket. Place the new base gasket over the crankcase studs and ease the barrel over the piston. When the barrel is in place turn the crankshaft to ensure smooth movement of the piston. Refit and tighten the four flange nuts over the holding studs. Place a new cylinder head gasket over the studs and replace the head. Tighten the nuts in a diagonal sequence, to a torque setting of 2.5 kgf m (18 lbf ft).
6   The exhaust pipe and carburettor hose may be fitted by reversing the dismantling sequence.

### 3 Crankcase covers: removal

**Right-hand cover**
1   Remove the single screw retaining the tachometer drive cable and pull the cable out of the cover. The kickstart lever can be pulled off its shaft after the Allen bolt has been removed.
2   The oil pump inspection cover is retained by three screws. Drain the transmission oil before removing the seven screws to release the main cover.
3   Before fitting the cover position a new gasket over the locating dowels.

**Left-hand cover**
4   Disconnect the gearchange linkage as described in Section 6. Remove the four screws to detach the inspection cover, so revealing the clutch operating mechanism. Disconnect the clutch cable from the operating arm and pull the cable out of the casing. The main cover can now be detached by removing its four screws.

### 4 Tachometer drive: cable removal

1   The tachometer drive cable is driven off a pinion mounted inside the right-hand crankcase cover. The cable can be pulled from the crankcase projection once the retaining screw has been removed.
2   Before refitting the cable check the housing O-ring and oil seal for signs of damage and leakage. The squared end of the cable must locate correctly in the corresponding housing of the drive gear, for the tachometer to function correctly.

### 5 Kickstart shaft: removal and refitting

1   Remove the Allen bolt from the kickstart lever knuckle and pull the lever off the shaft. Disconnect the tachometer cable and remove the right-hand crankcase cover as described in Sections 4 and 3.

**RS100 and 125 models**
2   Using a pair of long nose pliers unhook the return spring from its anchor pin. Grasp the shaft end firmly and pull it out of the crankcase.
3   To dismantle the assembly slide off the thick spacer and unclip the circlip to allow removal of the return spring and inner guide. Removal of the large plain washer will reveal the pinion retaining circlip, which when removed, will allow the collets and pinion to be slid off the shaft.

**RXS100 model**
4   Unhook the return spring ends and slide it and the inner guide off the shaft. The shaft can now be pulled out of the crankcase. Lift off the large plain washer and separate the pinion from the shaft.

**All models**
5   The kickstart mechanism is a robust assembly which should not normally require attention. Apart from obvious defects such as a broken return spring, the friction clip is the only component likely to cause problems if it becomes worn or weakened.
6   The friction clip can be checked using a spring balance. Hook one end of the balance onto the looped end of the friction clip. Pull on the free end of the balance and note the reading at the point where pressure overcomes the clip's resistance. This should normally be 0.8 – 1.3 kg (1.76 – 2.66 lb). If the reading is higher or lower than this and the mechanism has been malfunctioning, renew the clip as a precaution. Do not attempt to adjust a worn clip by bending it.
7   Refitting of the kickstart assembly can be conducted in a reverse order of the dismantling procedure.

### 6 Gearchange linkage: removal and refitting

1   Later models are fitted with a rearset type gearchange linkage. The main gearchange shaft splined end is clamped in an end-piece which is connected to the pedal via a linkage arm.
2   The complete assembly may be removed without dismantling the linkage by removing the end piece pinch bolt and the pedal circlip and washer. In order to maintain the present pedal height scribe a thin index line on the gearchange shaft end and end piece. The assembly can now be withdrawn. To dismantle the linkage, straighten and pull out the split pin to allow removal of the clevis pins.
3   Lubricate the linkage pivots before reassembly. When refitting the end piece to the splined shaft end line up the marks made when dismantling to preserve pedal height.

# Chapter 7 The 1977 on models

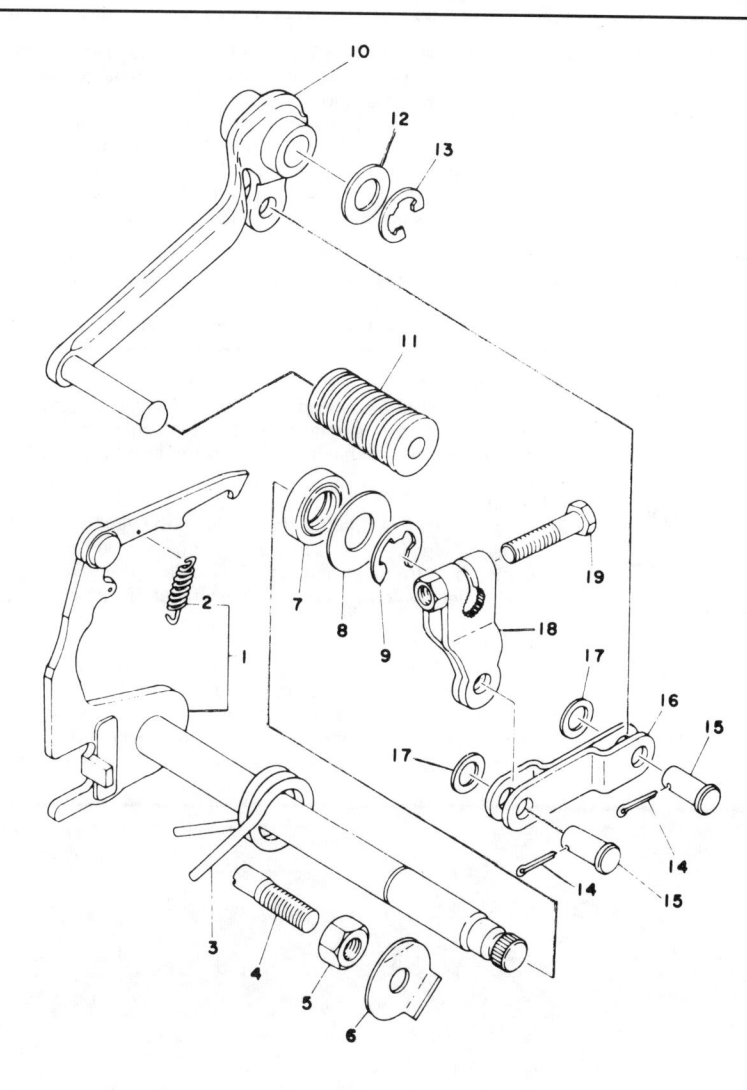

**Fig. 7.1 Gearchange linkage**

1. Gearchange shaft
2. Spring
3. Return spring
4. Return spring anchor
5. Nut
6. Tab washer
7. Oil seal
8. Washer
9. E-clip
10. Gearchange pedal
11. Pedal rubber
12. Washer
13. Circlip
14. Split pin – 2 off
15. Clevis pin – 2 off
16. Linkage arm – RS100
17. Washer – 2 off
18. End piece
19. Bolt
20. Linkage arm – RS125

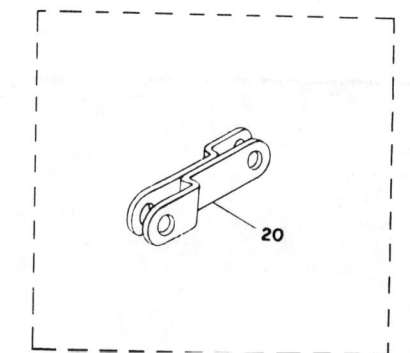

## 7 Crankshaft assembly: modification

1  Examination and renovation procedures remain unchanged from those given in Chapter 1, Section 16. Note that all models produced in 1977 onwards do not have a thrust washer fitted between the main bearings and the flywheels.

## 8 Gearbox layshaft: modification – RXS100

1  The layshaft 5th gear pinion is retained to its shaft by a circlip. Previous models have a thrust washer fitted between the pinion and circlip, as shown in photo 25.2c of Chapter 1. This thrust washer should not be fitted to RXS100 models.

## 9 Gear selector forks: modification – RXS100

1  The selector fork guide pin is now integral with the fork and can no longer be removed for renewal. If the pin is found to be worn beyond service limits then the complete fork will require renewal.
2  The forks are no longer located on their shafts by circlips but rely on the gear pinions and selector drum for location.

## 10 Petrol tap: removal and refitting – RXS100

1  The most likely cause of tap failure will be a blocked filter. To remove the filter, turn the tap lever to the 'Off' position and unscrew the bowl at the base of the tap. Pick out the O-ring and filter for inspection. The filter is a gauze element which can easily be cleaned by washing in clean petrol. If the filter or O-ring show signs of breakage they must be renewed.
2  If leaking is noticed from the operating lever, it can usually be attributed to O-ring failure. Disconnect the fuel feed pipe at the carburettor and place this end in a suitably sized metal can. Turn the tap to the 'Reserve position' and allow all the fuel to drain. Remove the two screws retaining the lever plate to allow removal of the lever, O-ring and spring.
3  In the event of the tap to tank sealing ring failing, remove the two screws and washers from the base of the tap. If these screws are difficult to slacken with the tank on the machine, remove the tank for better access. Withdraw the tap and inspect the sealing ring around it base.
4  Conduct tap reassembly in clean conditions, using only petrol for cleaning. Ensure that a fluid tight seal is made at every connection. Never use any of the proprietary silicone sealing compounds on the fuel system components. Petrol will cause the compound to break up and eventually cause carburettor jet blockage.

## 11 Carburettor: modifications – RS100 and 125

### RS100
1  The jet needle is now retained inside the throttle valve by a spring clip. Unscrew the carburettor top, compress the large return spring and disconnect the throttle cable trunnion from the base of the throttle valve. Using a small screwdriver hook out the W-shaped spring clip to permit needle removal.
2  The drain plug is situated on the side of the float chamber and takes the form of a cross-head screw. The sealing washer should be in good condition to prevent leakage.

### RS125
3  The drain plug is a hexagon head bolt screwed into the base of the float chamber. The sealing washer should be in good condition to prevent leakage.

### All models
4  The main jet can now be found in the top of the needle jet.

## 12 Carburettor: modifications – RXS100

1  The main jet screws into a brass holder, situated in the centre of the upturned carburettor. Remove both items to reveal the needle jet. This is a press fit in the body, but can be cleaned in place, by directing a blast of compressed air through its centre. Before refitting the main jet holder inspect the O-ring for damage. Refit and tighten the main jet carefully.
2  The drain plug is a hexagon head bolt screwed into the base of the float chamber. The sealing washer should be in good condition to prevent leakage.
3  Adjustment of the idle speed has been made easier by the fitting of a large knurled knob to the end of the throttle stop screw. Adjustment can now be made by hand.
4  The carburettor is connected to the reed valve adaptor by a rubber hose. This is simply held at each end by a worm drive clip.

## 13 Air cleaner: removal, cleaning and replacement

### RS100 and 125
1  Remove the left-hand side panel and the element cover, which is retained by a single screw. Pull the conical foam element out of the air cleaner box for examination and cleaning.

### RXS100
2  Remove the right-hand side panel and the element cover, which is retained by three screws. The flat foam element is held in position by a frame attached to the inside of the main cover. Carefully peel away the element for examination and cleaning.

### All models
3  A damaged element must be renewed immediately. Apart from the risk of damage from ingested dust, the holed filter will allow a much weaker mixture and may lead to overheating or seizure. To clean, wash the element thoroughly in petrol and when dry reimpregnate with oil. Squeeze carefully to remove

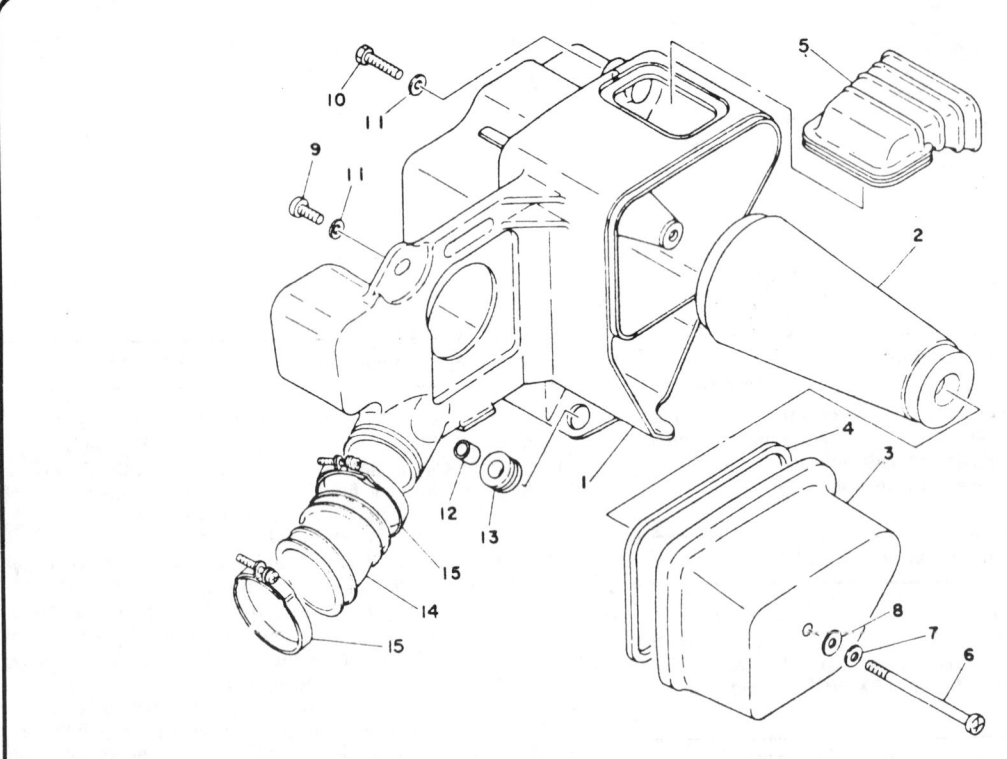

Fig. 7.2 Air cleaner – RS100 and 125

| 1 | Air cleaner box | 5 | Air intake hose | 9 | Screw – 2 off | 13 | Grommet |
| 2 | Foam element | 6 | Screw | 10 | Bolt | 14 | Air outlet hose |
| 3 | Element cover | 7 | Washer | 11 | Washer – 3 off | 15 | Hose clamp – 2 off |
| 4 | Sealing ring | 8 | Washer | 12 | Spacer | | |

Chapter 7 The 1977 on models 109

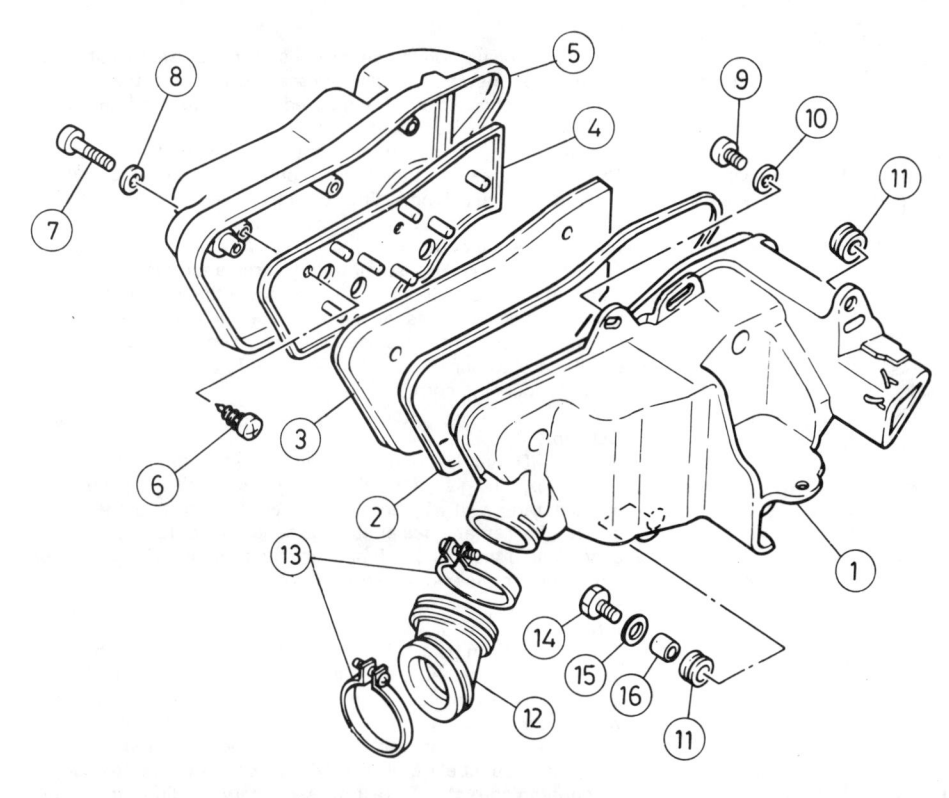

Fig. 7.3 Air cleaner – RXS100

1. Air cleaner box
2. Sealing ring
3. Foam element
4. Frame
5. Element cover
6. Screw – 2 off
7. Screw – 3 off
8. Washer – 3 off
9. Screw
10. Washer
11. Grommet
12. Air outlet hose
13. Hose clamp – 2 off
14. Bolt
15. Washer
16. Spacer

excess oil, do not wring out the element. The task of cleaning should be conducted at the following intervals if the machine is used in normal conditions.

RS100 and 125        Every 1000 miles (1500 km)
RXS100               Every 2000 miles (3000 km)

4  Reassemble the element, cover and side panel ensuring that the assembly is completely airtight.

## 14 YEIS unit: removal and refitting – RXS 100

1  The purpose of this device is to compensate for the variations in pressure created in the mixture path to the engine. The YEIS reservoir is mounted across the frame top tubes and connected to the inlet tract by a hose.
2  To remove, detach the petrol tank and remove the reservoir retaining screw and washer. Squeeze together the ears of the hose clip and pull the hose off the reed valve block union. The YEIS unit can now be lifted off the machine.
3  Examine the reservoir for cracks and splits and the hose for signs of perishing. If it is necessary to renew either component it is essential that the genuine Yamaha part is purchased in order for the system to function correctly.
4  Refitting is a direct reversal of the dismantling procedure but take care to ensure airtight hose connections are made.

## 15 Exhaust system: removal and refitting – RXS 100

1  The exhaust system comprises two parts, these being the front pipe and silencer.
2  Remove the two nuts at the cylinder barrel exhaust port flange, the silencer front mounting bolt and the single bolt retaining the rear mounting plate. The complete system may now be manoeuvred clear of the machine. If it is only necessary to remove the front pipe, for example when removing the cylinder barrel, simply remove the two nuts at the barrel flange and unscrew the joining gland nut.
3  Conduct removal and cleaning of the silencer baffle as described in Chapter 2, Section 13.
4  Before refitting the system check the condition of the sealing ring in the cylinder barrel port and the sealing ring at the gland nut join.

## 16 Oil pump: drive pinion removal

1  The removal and refitting procedure remains virtually unchanged from the information given in Chapter 2, Section 15. The only exception being the method of retaining the drive pinion.
2  The pinion is held on its shaft by a large E-clip. With the clip removed the pinion can be drawn off. When refitting ensure that the cut-outs in the rear face of the pinion align with the corresponding shaft pin.

## 17 Oil level gauge and warning lamp

1  A gauge is incorporated in the side of the oil tank for ease of checking the level. It is operated by a float inside the tank.
2  Removal of the gauge can be accomplished after disconnecting the oil feed pipe and draining the oil into a suitable container. Remove the three screws and spring washers to release the retaining plate and pull the gauge from position. When free, manoeuvre the float arm out of the tank.
3  When refitting the unit ensure a fluid tight seal is made around the gauge base. Reconnect the feed pipe to the base of the tank. Disconnect the other end of the pipe, where it meets the engine and allow oil to flow until the pipe is bled of air.
4  The warning lamp, situated in the speedometer, is operated by a float type switch which is a press fit in the top of the oil tank. This lamp will illuminate when the oil level in the tank is too low. The circuit is wired through the neutral switch so that

when the ignition is switched on and the machine is in neutral, the lamp comes on as a means of checking its operation. As soon as a gear is selected the lamp should go out unless the oil level is low.

5  In the event of a fault the bulb can be checked by switching the ignition on and selecting neutral. If this proves sound, check for 6 volts on the black/red lead to the switch. If the switch proves to be defective it can be unclipped from the tank and withdrawn.

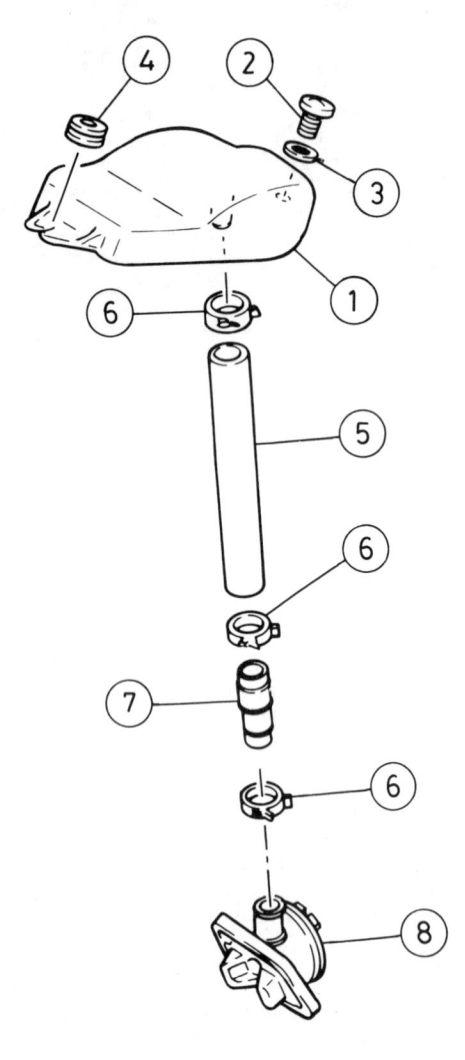

**Fig. 7.4 YEIS unit**

| 1 | YEIS reservoir | 5 | Hose |
| 2 | Screw | 6 | Hose clip |
| 3 | Washer | 7 | Union |
| 4 | Grommet | 8 | Reed valve block |

## 18 Electronic ignition system – RXS100

1  The advantage of this system over the conventional magneto and contact breaker arrangement is that it uses no mechanical components and hence, the need for constant maintenance and adjustment does not arise. Once the electronic system is set up it need not be attended to unless it has been disturbed in the course of dismantling or unless failure occurs in the components. The following tests can be conducted if component failure is suspected.

### Pulser coil
2  This is located on the alternator stator and can be tested by using a multimeter set on its ohm scale. Disconnect the plastic block connector and single white/red wire connector from the alternator wiring. Connect the multimeter across the black wire terminal of the block connector and the single white/red wire. The reading obtained should be 20 ohm ± 10% at 20°C (68°F), if the pulser coil is functioning correctly.

### Ignition source coil
3  The source coil is also mounted on the alternator stator, adjacent to the pulser coil. To test, connect the multimeter set on its ohm scale across the black wire terminal of the alternator block connector and the single black/red wire. The reading should be 240 ohm ± 10% at 20°C (68°F) if the ignition source coil is functioning correctly.

### CDI unit
4  The CDI unit is mounted on the frame top tubes and retained by a single screw. No test figures are given by the manufacturer so it will therefore be necessary to seek the advice of a Yamaha dealer as to its condition. Before condemning the CDI unit first check the terminals and wiring for loose connectors and frayed wires.

### Ignition coil
5  This test can be conducted with the unit on the machine, having first disconnected the high tension lead at the spark plug and the low tension lead at the connector. To test the primary windings, set the multimeter on its ohm scale and connect it across the orange wire and the coil earth pole. The reading obtained should be 1.6 ohm ± 10% @ 20°C (68°F) if the coil is functioning correctly. To test the secondary windings, move the

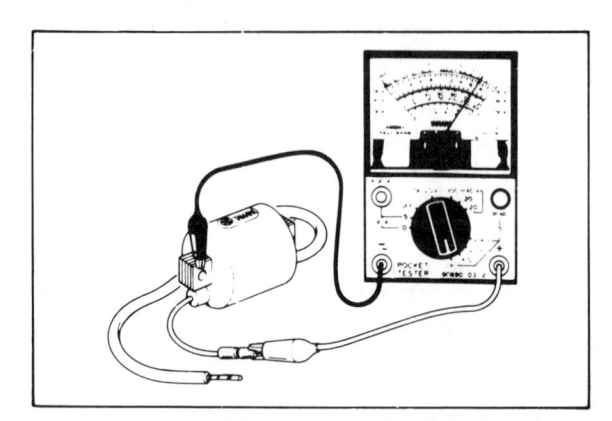

Primary coil test

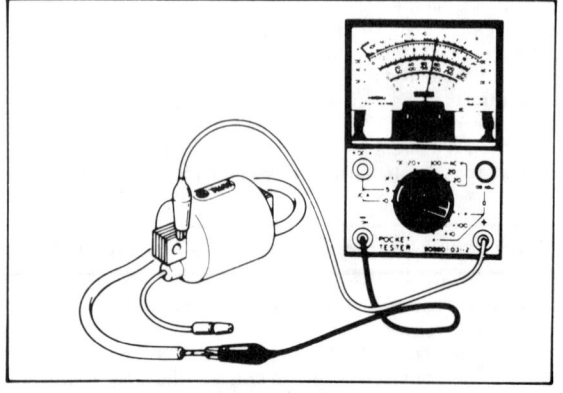

Secondary coil test

**Fig. 7.5 Ignition coil resistance tests**

# Chapter 7 The 1977 on models

multimeter to its K ohm function and connect it across the HT lead and the coil earth pole. The reading obtained should be 6.6 K ohm ± 20% @ 20°C (68°F).

**Ignition timing check**

6  The ignition timing on this model is not adjustable, neither is there means of checking the timing. If the timing is suspected to be at fault, test the other electronic components first, to check that the fault does not lie elsewhere in the system.

## 19 Steering head: modifications

1  The handlebars are retained to the top yoke by two separate clamps. Each clamp being held by two bolts. When refitting the handlebars position the knurled sections exactly under each clamp top half. The torque setting for these bolts is 2.0 kgf m (14 lbf ft).
2  The fork legs are held in the top yoke by a pinch-bolt on each side. The torque setting for this bolt is 1.0 kgf m (7.5 lbf ft) for RS100 and 125 models and 1.5 kgf m (11 lbf ft) for RXS100 models.

**RXS100 models only**

3  If removal of the top yoke is necessary it will be found that a spacer is fitted over the bearing adjusting nut.
4  An emblem plate is fitted across the lower yoke. This can be removed by unscrewing the two screws and washers.

## 20 Front fork legs: dismantling and reassembly

1  Support the machine on a suitable stand and remove the front wheel and mudguard, and brake caliper (where fitted). These operations are described in Chapter 5.
2  Using a small screwdriver prise out the rubber cap from the top of each leg. The top plug is retained by a circlip on RS100 and 125 models and when removed, the plug will be expelled by spring pressure. The top plug fitted to the RXS100 is threaded and screws into the stanchion. This can be removed with an appropriate size Allen key.
3  Slacken the upper and lower yoke pinch bolts and pull each leg downwards, out of the headlamp brackets and steering head. Dismantle each leg separately to avoid interchanging components.
4  With the top plug removed, invert the leg and allow the oil, spacer, spring seat and spring to be removed, pumping the leg to expel all the oil. To separate the stanchion from the lower leg remove the Allen bolt from the base of the assembly. Depending on which model is being worked on the wheel spindle clamp may have to be taken off to gain access to the bolt. The Allen bolt screws into the bottom of the damper rod which is free to revolve inside the stanchion. It is often found that some means of holding the rod is necessary in order to slacken the bolt. To overcome this Yamaha have produced a service tool (Part No 90890-01212) which is essentially a long rod with a shaped head which locates in the damper rod head. A home made equivalent can be made from a length of wooden dowelling. The tool must be at least the length of the fork leg and have one end slightly tapered. Pass the tapered end down the stanchion and press it into the damper rod head. With the help of an assistant the tool can be held firm whilst the Allen bolt is slackened. Remove the Allen bolt and sealing washer and separate the stanchion from the lower leg. The damper rod and rebound spring can be shaken out of the stanchion. Note that the RXS100 is not fitted with a rebound spring.
5  Slide the rubber dust seal off the stanchion and inspect it for splitting or cracking. This seal must be in good condition, because it forms an important function of stopping water and grit from damaging the oil seal. Inspect the stanchion and lower leg for signs of pitting and scoring. Pitting of the stanchion is caused by road chips and corrosion and can be renovated by filling the indentations with Araldite. This repair will only be successful if the stanchion is absolutely clean and the compound is applied smoothly. An old oil seal is ideal for this purpose.
6  The lower leg oil seal is retained by a wire clip. Prise out the clip and, using a large screwdriver, lever out the oil seal. Damage will almost certainly be caused to the oil seal upon removal, therefore it must be renewed. Yamaha also recommend renewal of the wire clip. When fitting the new oil seal ensure that it enters its housing squarely when driving it into place with a block of wood or suitable size socket.
7  Before reassembly, ensure that all components are clean and use only the recommended fork oil to lubricate the oil seal lips and stanchion. Drop the damper rod seat into the lower leg and insert the rebound spring (where fitted) and damper rod into the stanchion. Pass the stanchion through the oil seal and into the lower leg. Position the sealing washer over the Allen bolt and screw the bolt into the base of the lower leg. The tool used when dismantling can now be used for tightening the Allen bolt. Slide the rubber dust seal over the stanchion and fit it over the lower leg.
8  Using the recommended grade of oil fill the leg to the capacity given in the Specifications Section. Insert the spring, spring seat and spacer. The leg can now be positioned in the yokes and the pinch bolts tightened to the recommended torque setting. Fit the top plug assembly and the rubber cap.
9  Refit the front wheel, mudguard and brake caliper and pump the forks up and down to check for correct operation.

## 21 Swinging arm: pivot bolt – RS100 1981 and 1982

1  The only change made to the swinging arm is the direction of fitting the pivot bolt. It passes through the assembly from the left-hand side.
2  Before tightening the securing nut a washer must be positioned each side of the exhaust mounting bracket.

## 22 Swinging arm: bush removal – RXS100

1  Follow the instructions given in Chapter 4, Section 12 for swinging arm removal from the machine.
2  With the assembly positioned on a workbench pull off the dust cover and shim from the outer face of each pivot boss. The bush comprises two parts, the fibre bush, which is a press fit in the boss and an inner metal sleeve. The metal sleeve should be able to be pushed out using finger pressure. Prise out the oil seal from the inner face of each boss. This leaves the fibre bush which must be driven out by judicious use of a hammer and drift. The bush is shouldered at one end and can only be driven outwards from the boss. It is suggested that bush removal should only be contemplated if a new bushes are to be fitted. This is because they are made of a very brittle material, which will probably fracture when being subjected to force during removal.
3  Before fitting new bushes remove any trace of corrosion from the swinging arm boss and lightly grease the surfaces to aid bush insertion. Ensure that the new bush enters its housing squarely. To prevent chipping use a soft wood or hard rubber pad between the bush and hammer. Grease the metal sleeve before inserting in the bush. If the inner oil seal was damaged during dismantling it must be renewed. It can be refitted in the same manner as the bush. Finally position the shim and dust cover over the pivot boss ends.
4  The swinging arm may be refitted to the machine by reversing the dismantling procedure. Grease the pivot shaft with high melting point grease before insertion and check that the arm pivots smoothly, without binding. Refit the pivot bolt washer and nut, tightening the nut to 5.8 kgf m (42 lbf ft).

## Chapter 7 The 1977 on models

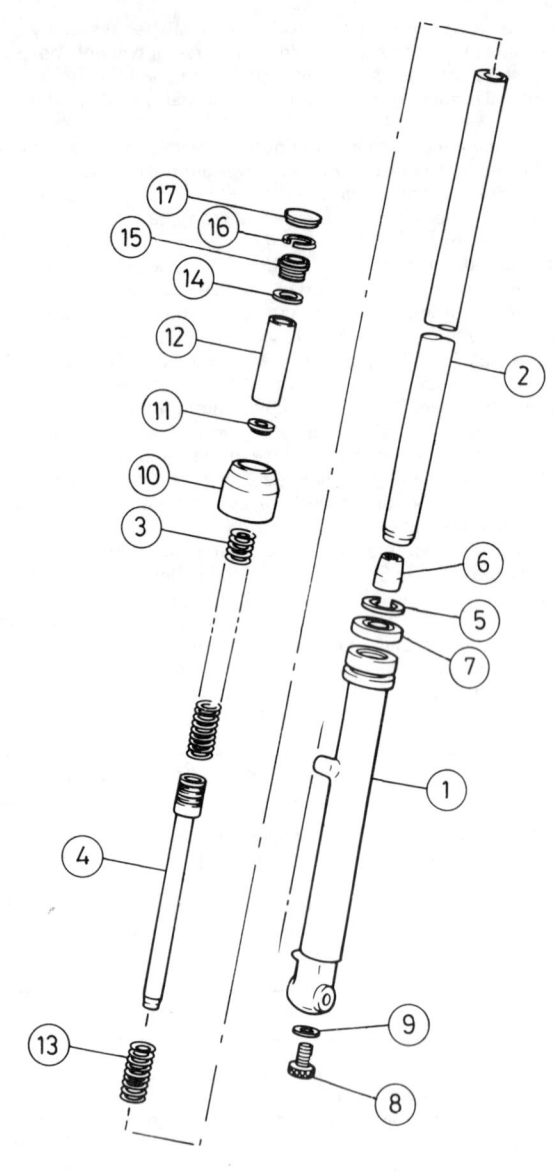

Fig. 7.6 Front forks

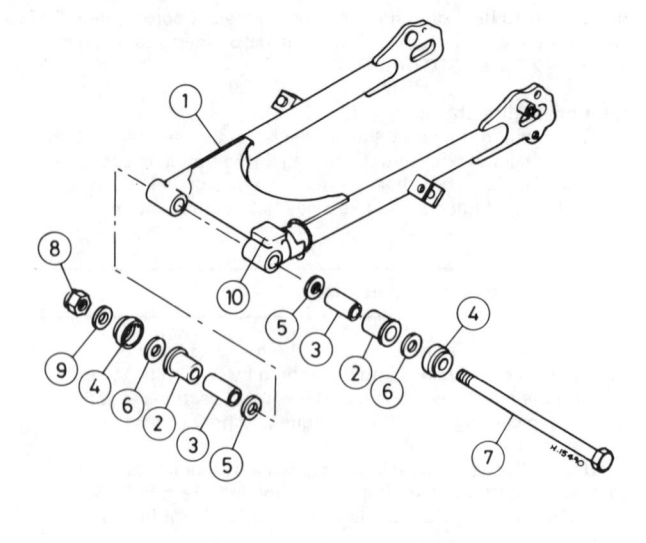

1  Lower leg
2  Stanchion
3  Spring
4  Damper rod
5  Wire circlip
6  Damper rod seat
7  Oil seal
8  Allen bolt
9  Sealing washer
10 Dust seal
11 Spring seat
12 Spacer
13 Rebound spring – RS models only
14 O-ring
15 Top plug
16 Circlip – RS models only
17 Rubber cap

Fig. 7.7 Swinging arm – RXS100

1  Swinging arm
2  Bush
3  Metal sleeve
4  Dust cover
5  Oil seal
6  Shim
7  Pivot shaft
8  Nut
9  Washer

### 23 Front brake caliper: pad renewal – RS125 1981 on

1  Remove the two bolts retaining the caliper to the fork leg and lift the caliper off the disc. Each pad is held in an anti-rattle shim by a thin spring clip. Prise out the clip and pick out the pad. The moving pad may prove a little difficult to remove. If so, the brake lever may be operated a few times to push the pad from position. Take care when doing this not to force out completely the piston from the caliper body.

2  Clean away all traces of dirt from the pads and inspect them for chipping and breakage. If the braking surface is glazed rub with emery paper to roughen the surface and break the glaze. A wear line should be visible around the edge of the pad. When the friction material wears down to this line the pads must be renewed.

3  Refit the anti-rattle spring, pad and spring clip to the caliper. If the piston pressure was used to force the moving pad from position the piston must be pushed back in the caliper. Firm thumb pressure should suffice for this purpose. Replace the caliper on the disc and refit the mounting bolts, with their spring and plain washers.

### 24 Front brake caliper: dismantling and reassembly – RS125 1981 on

1  Before the caliper is dismantled wash the outer surfaces to remove as much road grime as possible. Also have ready a new can of brake fluid. Remove the caliper and brake pads as detailed in the previous section.

2  Before disconnecting the hydraulic hose use the pressure to push the piston out of its bore. Obtain a large, clean, polythene bag. Put the caliper in the bag and hold the neck of the bag closed around the hose. Repeatedly apply the brake lever to push the piston out of the bore, into the bag. Remove the caliper from the bag and allow the remaining fluid to drain, by removing the brake hose banjo union bolt. Wipe away the remaining traces of fluid and inspect the caliper.

3  Pick out the two piston seals from the caliper bore and inspect them for damage. Clean the piston and seals in new brake fluid. If fluid leakage has been noticed and the seals appear to be in good condition check the piston and bore for scoring. The piston and seals cannot be purchased separately, they must therefore be treated very carefully if unnecessary expense is to be avoided.

4  Refit the seals into their caliper locations and push in the piston. Coating the piston surface with brake fluid will aid reassembly. The brake pads may be refitted as described in the previous section.

## Chapter 7 The 1977 on models

5   The caliper pivot bolt passes through the back of the body and may often become stiff due to inadequate lubrication. To remove, straighten the split pin and undo the nut and washer. Pull out the bolt from the top of the caliper. Use fine emery paper to remove any corrosion build up on the shaft and caliper bore. Before inserting the bolt apply a film of high melting-point grease to its surface.
6   Mount the caliper on the disc and refit the two fork leg bolts. Reconnect the hydraulic brake hose. Remove the master cylinder cap and fill the system with new fluid. Follow the procedure given in Chapter 5, Section 10 for bleeding the brake.

### 25 Rear wheel: modifications – RXS 100

1   The rear wheel removal and refitting procedure remains virtually unchanged from the instructions given in Chapter 5, Sections 17-20, with the following exceptions.
2   The wheel spindle passes through the assembly from the left-hand side.
3   The right-hand wheel bearing no longer has an oil seal fitted.
4   When removing the cush drive assembly note that the bearing is no longer retained by a large circlip. Also the dust seal preceding the bearing has been replaced by an oil seal.

### 26 Headlamp: bulb renewal – RXS100

1   Remove the two screws from the underside of the unit to release the headlamp from its shell. Gently lift out the rim and reflector far enough for the block connector to be unplugged from the bulb.
2   Depress the large bulbholder ring against spring pressure and rotate it anti-clockwise to release the bulb.
3   The pilot bulb is of the bayonet fitting type and is situated in the base of the reflector. This can simply be pulled from position.

### 27 Speedometer and tachometer heads: bulb renewal

1   To facilitate instrument bulb removal the instrument must be raised from the main mounting bracket. First unscrew the knurled ring around the drive cable and pull the cable clear of the instrument head. Each instrument is retained by two nuts under the mounting bracket. Raise the instrument and unplug the affected bulb holder.

2   The warning lamp bulbs are contained in a separate compartment between the speedometer and tachometer. To release the top cover remove the two retaining screws. On RXS100 models the screws can be found under the mounting bracket, but on all other models they are positioned on top.
3   When refitting either instrument ensure that the damping rubbers are fitted correctly.

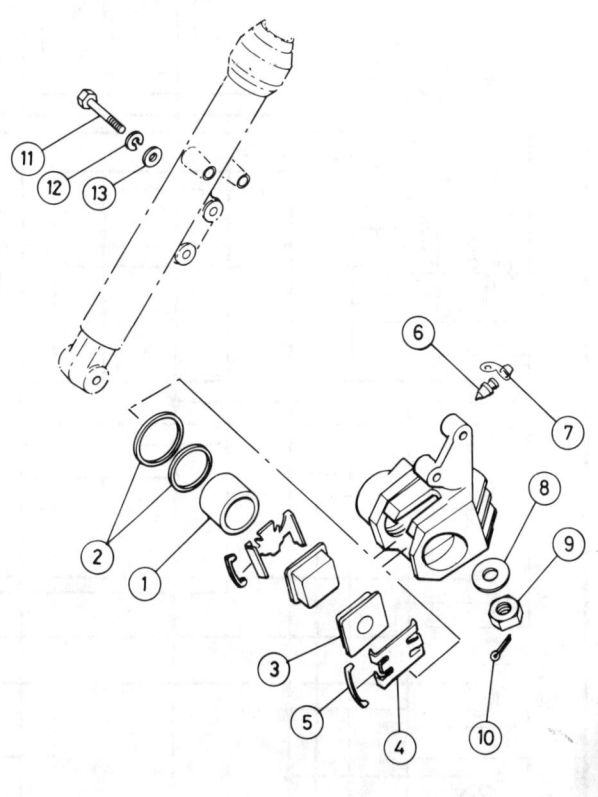

Fig. 7.8 Front brake caliper – RS125 1981 to 1984

1   Piston
2   Piston seals
3   Brake pad – 2 off
4   Anti-rattle shim – 2 off
5   Spring clip – 2 off
6   Bleed nipple
7   Bleed nipple cap
8   Washer
9   Nut
10  Split pin
11  Bolt – 2 off
12  Spring washer – 2 off
13  Washer – 2 off

Wiring diagram – RS100, RS125 and RS125 DX 1974 to 1976 models

See page 119 for key

Wiring diagram – RS100 1977 model onwards

*See page 119 for key*

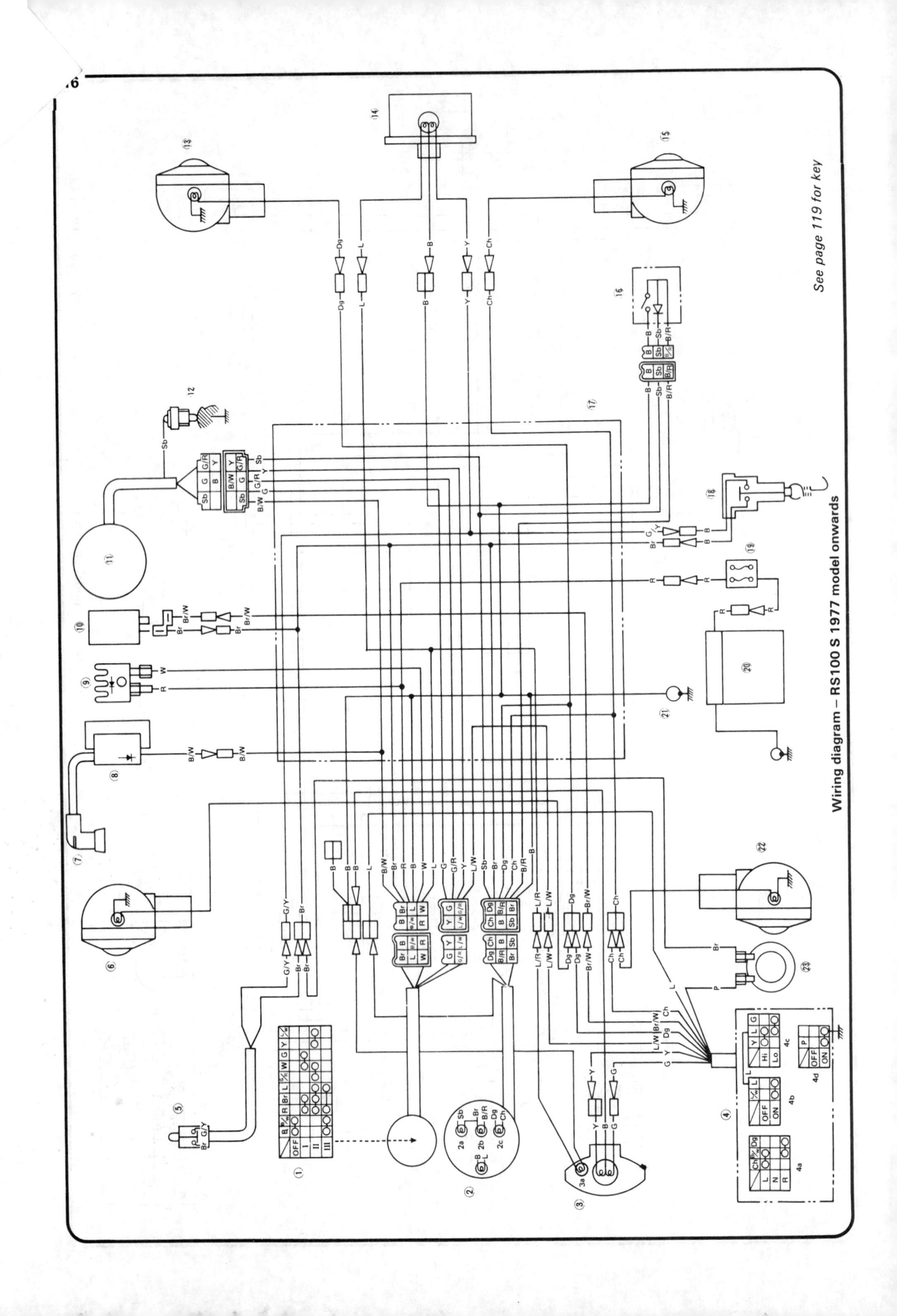

Wiring diagram – RS100 S 1977 model onwards

See page 119 for key

117

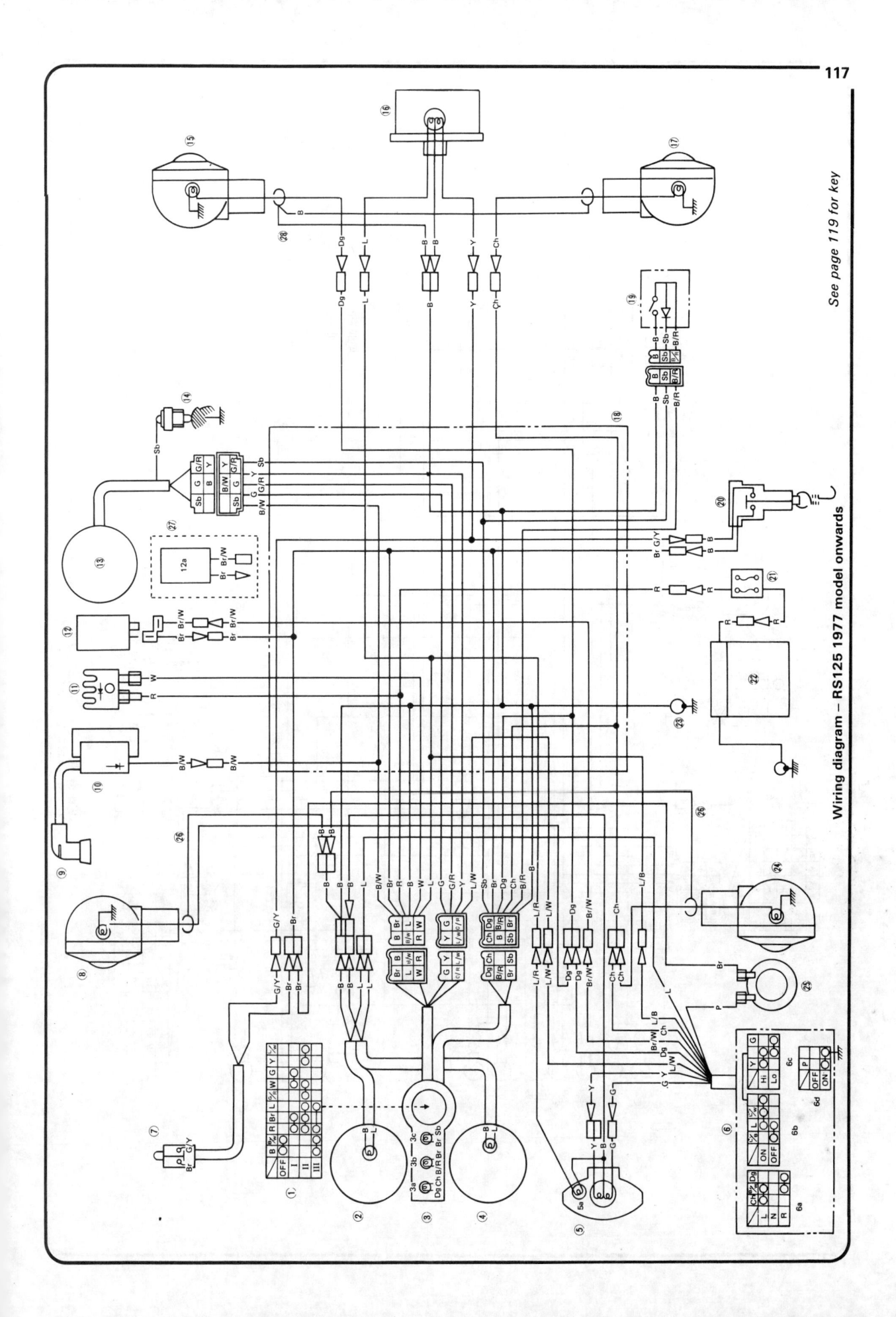

Wiring diagram – RS125 1977 model onwards

*See page 119 for key*

Wiring diagram – RXS100 1983 model onwards

See page 119 for key

### Key to RS100 wiring diagram (1977 on models)

- 1 Ignition/lighting switch
- 2 Tachometer
- 3 Indicator/warning lamps
- 3a Flashing indicator warning lamp
- 3b Oil tank level warning lamp
- 3c Neutral indicator lamp
- 4 Speedometer
- 5 Headlamp unit
- 5a Pilot light
- 6 Handlebar switch unit
- 6a Flashing indicator switch
- 6b Lighting switch
- 6c Dip switch
- 6d Horn button
- 7 Front brake light switch
- 8 RH front indicator
- 9 Spark plug cap
- 10 Ignition coil
- 11 Rectifier
- 12 Flasher relay
- 13 Flywheel generator
- 14 Neutral indicator switch
- 15 RH rear indicator
- 16 Stop/tail light
- 17 LH rear indicator
- 18 Wiring harness
- 19 Oil level warning sender
- 20 Rear brake light switch
- 21 Fuse
- 22 Battery
- 23 Earth
- 24 LH front indicator
- 25 Horn
- 26 Germany only
- 27 Belgium only

### Key to RS100 S wiring diagram (1977 on models)

- 1 Ignition/lighting switch
- 2 Speedometer
- 2a Neutral indicator light
- 2b Oil level warning light
- 2c Indicator warning light
- 3 Headlamp unit
- 3a Pilot light
- 4 Handlebar switch unit
- 4a Indicator switch
- 4b Lighting switch
- 4c Dip switch
- 4d Horn button
- 5 Front brake light switch
- 6 RH front indicator
- 7 Spark plug cap
- 8 Ignition coil
- 9 Rectifier
- 10 Flasher relay
- 11 Flywheel generator
- 12 Neutral indicator switch
- 13 RH rear indicator
- 14 Stop/tail light
- 15 LH rear indicator
- 16 Oil level warning sender
- 17 Wiring harness
- 18 Rear brake light switch
- 19 Fuse
- 20 Battery
- 21 Earth
- 22 LH front indicator
- 23 Horn

### Key to RS125 wiring diagram (1977 on models)

- 1 Ignition/lighting switch
- 2 Tachometer
- 3 Indicator/warning lamps
- 3a Indicator warning light
- 3b Oil level warning lamp
- 3c Neutral indicator lamp
- 4 Speedometer
- 5 Headlamp unit
- 5a Pilot light
- 6 Handlebar switch unit
- 6a Indicator switch
- 6b Lighting switch
- 6c Dip switch
- 6d Horn button
- 7 Front brake light switch
- 8 RH front indicator
- 9 Spark plug cap
- 10 Ignition coil
- 11 Rectifier
- 12 Flasher relay
- 13 Flywheel generator
- 14 Neutral indicator switch
- 15 RH rear indicator
- 16 Stop/tail light
- 17 LH rear indicator
- 18 Wiring harness
- 19 Oil level warning sender
- 20 Rear brake light switch
- 21 Fuse
- 22 Battery
- 23 Earth
- 24 LH front flasher
- 25 Horn
- 26 Excluding Belgian models
- 27 Switzerland, Finland only
- 28 Switzerland, Finland only

### Key to RXS 100 wiring diagram (1983 on models)

- 1 Ignition switch
- 2 Front brake light switch
- 3 Ignition coil
- 4 CDI unit
- 5 Voltage regulator
- 6 Rectifier
- 7 Flywheel generator
- 8 Flasher relay
- 9 Oil level switch
- 10 Rear indicators
- 11 Stop/tail lamp
- 12 Rear brake light switch
- 13 Fuse
- 14 Battery
- 15 Earth point
- 16 Horn
- 17 Horn button
- 18 Indicator switch
- 19 Lighting switch
- 20 Headlamp dip switch
- 21 Auxiliary lamp
- 22 Front indicators
- 23 Headlamp
- 24 Instrument console lamp
- 25 Indicator warning lamp
- 26 High beam warning lamp
- 27 Oil level warning lamp
- 28 Neutral position lamp

### Key to RS100, RS125 and RS125 DX wiring diagram (1974 to 1976 models)

- 1 Fuse
- 2 Rear brake light switch
- 3 Front brake light switch
- 4 Neutral indicator lamp
- 5 Flasher relay
- 6 Horn
- 7 Tachometer (RS125 DX only)
- 8 Speedometer
- 9 Ignition/lighting switch
- 10 RH rear indicator
- 11 Stop/tail light
- 12 LH rear indicator
- 13 Horn button
- 14 Lighting switch
- 15 Dip switch
- 16 Indicator switch
- 17 Flywheel generator
- 18 Rectifier
- 19 Resistor
- 20 Ignition coil
- 21 Spark plug
- 22 Earth
- 23 LH front indicator
- 24 Pilot light
- 25 Headlamp unit
- 26 RH front indicator
- 27 Battery
- 28 Instrument lamp (RS125 DX only)
- 29 Neutral indicator light
- 30 Instrument lamp
- 31 Indicator warning lamp

### Wiring diagrams – colour key

- Y   Yellow
- Br  Light brown
- G/Y Green/Yellow
- R   Red
- L   Blue
- B/W Black/White
- G/R Green/Red
- B/R Black/Red
- L/W Blue/White
- P   Pink
- Ch  Dark brown
- Br/W Light brown/White
- Dg  Dark green
- B   Black
- G   Green
- Gr  Grey
- O   Orange
- W   White

# Conversion factors

*Length (distance)*
| | | | | | |
|---|---|---|---|---|---|
| Inches (in) | X | 25.4 | = Millimetres (mm) | X 0.0394 | = Inches (in) |
| Feet (ft) | X | 0.305 | = Metres (m) | X 3.281 | = Feet (ft) |
| Miles | X | 1.609 | = Kilometres (km) | X 0.621 | = Miles |

*Volume (capacity)*
| | | | | | |
|---|---|---|---|---|---|
| Cubic inches (cu in; in³) | X | 16.387 | = Cubic centimetres (cc; cm³) | X 0.061 | = Cubic inches (cu in; in³) |
| Imperial pints (Imp pt) | X | 0.568 | = Litres (l) | X 1.76 | = Imperial pints (Imp pt) |
| Imperial quarts (Imp qt) | X | 1.137 | = Litres (l) | X 0.88 | = Imperial quarts (Imp qt) |
| Imperial quarts (Imp qt) | X | 1.201 | = US quarts (US qt) | X 0.833 | = Imperial quarts (Imp qt) |
| US quarts (US qt) | X | 0.946 | = Litres (l) | X 1.057 | = US quarts (US qt) |
| Imperial gallons (Imp gal) | X | 4.546 | = Litres (l) | X 0.22 | = Imperial gallons (Imp gal) |
| Imperial gallons (Imp gal) | X | 1.201 | = US gallons (US gal) | X 0.833 | = Imperial gallons (Imp gal) |
| US gallons (US gal) | X | 3.785 | = Litres (l) | X 0.264 | = US gallons (US gal) |

*Mass (weight)*
| | | | | | |
|---|---|---|---|---|---|
| Ounces (oz) | X | 28.35 | = Grams (g) | X 0.035 | = Ounces (oz) |
| Pounds (lb) | X | 0.454 | = Kilograms (kg) | X 2.205 | = Pounds (lb) |

*Force*
| | | | | | |
|---|---|---|---|---|---|
| Ounces-force (ozf; oz) | X | 0.278 | = Newtons (N) | X 3.6 | = Ounces-force (ozf; oz) |
| Pounds-force (lbf; lb) | X | 4.448 | = Newtons (N) | X 0.225 | = Pounds-force (lbf; lb) |
| Newtons (N) | X | 0.1 | = Kilograms-force (kgf; kg) | X 9.81 | = Newtons (N) |

*Pressure*
| | | | | | |
|---|---|---|---|---|---|
| Pounds-force per square inch (psi; lbf/in²; lb/in²) | X | 0.070 | = Kilograms-force per square centimetre (kgf/cm²; kg/cm²) | X 14.223 | = Pounds-force per square inch (psi; lbf/in²; lb/in²) |
| Pounds-force per square inch (psi; lbf/in²; lb/in²) | X | 0.068 | = Atmospheres (atm) | X 14.696 | = Pounds-force per square inch (psi; lbf/in²; lb/in²) |
| Pounds-force per square inch (psi; lbf/in²; lb/in²) | X | 0.069 | = Bars | X 14.5 | = Pounds-force per square inch (psi; lbf/in²; lb/in²) |
| Pounds-force per square inch (psi; lbf/in²; lb/in²) | X | 6.895 | = Kilopascals (kPa) | X 0.145 | = Pounds-force per square inch (psi; lbf/in²; lb/in²) |
| Kilopascals (kPa) | X | 0.01 | = Kilograms-force per square centimetre (kgf/cm²; kg/cm²) | X 98.1 | = Kilopascals (kPa) |
| Millibar (mbar) | X | 100 | = Pascals (Pa) | X 0.01 | = Millibar (mbar) |
| Millibar (mbar) | X | 0.0145 | = Pounds-force per square inch (psi; lbf/in²; lb/in²) | X 68.947 | = Millibar (mbar) |
| Millibar (mbar) | X | 0.75 | = Millimetres of mercury (mmHg) | X 1.333 | = Millibar (mbar) |
| Millibar (mbar) | X | 0.401 | = Inches of water (inH$_2$O) | X 2.491 | = Millibar (mbar) |
| Millimetres of mercury (mmHg) | X | 0.535 | = Inches of water (inH$_2$O) | X 1.868 | = Millimetres of mercury (mmHg) |
| Inches of water (inH$_2$O) | X | 0.036 | = Pounds-force per square inch (psi; lbf/in²; lb/in²) | X 27.68 | = Inches of water (inH$_2$O) |

*Torque (moment of force)*
| | | | | | |
|---|---|---|---|---|---|
| Pounds-force inches (lbf in; lb in) | X | 1.152 | = Kilograms-force centimetre (kgf cm; kg cm) | X 0.868 | = Pounds-force inches (lbf in; lb in) |
| Pounds-force inches (lbf in; lb in) | X | 0.113 | = Newton metres (Nm) | X 8.85 | = Pounds-force inches (lbf in; lb in) |
| Pounds-force inches (lbf in; lb in) | X | 0.083 | = Pounds-force feet (lbf ft; lb ft) | X 12 | = Pounds-force inches (lbf in; lb in) |
| Pounds-force feet (lbf ft; lb ft) | X | 0.138 | = Kilograms-force metres (kgf m; kg m) | X 7.233 | = Pounds-force feet (lbf ft; lb ft) |
| Pounds-force feet (lbf ft; lb ft) | X | 1.356 | = Newton metres (Nm) | X 0.738 | = Pounds-force feet (lbf ft; lb ft) |
| Newton metres (Nm) | X | 0.102 | = Kilograms-force metres (kgf m; kg m) | X 9.804 | = Newton metres (Nm) |

*Power*
| | | | | | |
|---|---|---|---|---|---|
| Horsepower (hp) | X | 745.7 | = Watts (W) | X 0.0013 | = Horsepower (hp) |

*Velocity (speed)*
| | | | | | |
|---|---|---|---|---|---|
| Miles per hour (miles/hr; mph) | X | 1.609 | = Kilometres per hour (km/hr; kph) | X 0.621 | = Miles per hour (miles/hr; mph) |

*Fuel consumption**
| | | | | | |
|---|---|---|---|---|---|
| Miles per gallon, Imperial (mpg) | X | 0.354 | = Kilometres per litre (km/l) | X 2.825 | = Miles per gallon, Imperial (mpg) |
| Miles per gallon, US (mpg) | X | 0.425 | = Kilometres per litre (km/l) | X 2.352 | = Miles per gallon, US (mpg) |

*Temperature*

Degrees Fahrenheit = (°C x 1.8) + 32        Degrees Celsius (Degrees Centigrade; °C) = (°F - 32) x 0.56

*It is common practice to convert from miles per gallon (mpg) to litres/100 kilometres (l/100km), where mpg (Imperial) x l/100 km = 282 and mpg (US) x l/100 km = 235

# Index

## A

**Acknowledgements** 2
**Adjustment:-**
   brake 9, 84, 88
   carburettor 48
   clutch 11
   contact breaker points 9, 56
   final drive chain 8, 90
   headlamp beam 95
   ignition timing 9, 57
   oil pump 9, 41, 54
   spark plug 60
   wheel alignment 90
**Air filter** 9, 50, 108

## B

**Battery** 8, 94
**Bearings:-**
   engine 25
   steering head 62
   wheel 11
**Bleeding:-**
   hydraulic brake 84
   oil pump 41, 53
**Brakes:-**
   drum-type:
      adjustment 9, 84, 88
      examination and renovation 78
      wear check 11
   fault diagnosis 92
   hydraulic-type:
      bleeding 84
      caliper 80, 112
      examination and renovation 78
      fluid level check 7
      master cylinder 83
      pad renewal 80, 112
      pad wear 11
   specifications 76, 104

**Bulbs:-**
   flashing indicators 96
   headlamp 95
   instruments 97, 113
   stop and tail lamp 96
   wattages 93, 105

## C

**Cables:-**
   clutch 11, 41
   front brake 84
   instrument 73, 106
   lubrication 8
   oil pump 41, 54
   throttle 46, 54
**Carburettor:-**
   dismantling and examination 45
   modifications 108
   reassembly 46
   removal 45
   settings 48
   specifications 44
**CDI unit** 108
**Centre stand** 70
**Chain – final drive** 8, 90
**Clutch:-**
   adjustment 11
   cable 11, 41
   examination and renovation 28
   fault diagnosis 43
   refitting 35
   removal 21
   specifications 15, 103
**Condenser** 58
**Contact breaker** 9, 56
**Conversion factors** 120
**Crankcases:-**
   examination and renovation 33
   joining 35
   separating 22

# Index

**Crankshaft:-**
    examination and renovation 25
    modifications 107
    removal 22
**Cush drive** 88
**Cylinder head and barrel:-**
    examination and renovation 27
    refitting 40, 106
    removal 19, 106

## D

**Decarbonising:-**
    engine 11, 27
    exhaust 9, 52
**Dualseat** 73

## E

**Electrical system:-**
    battery 8, 94
    charging system 94
    fault diagnosis 100
    flashing indicators 96, 97
    fuse 95
    handlebar switches 96
    headlamp 95, 113
    horn 98
    instruments 97
    rectifier 94
    specifications 93
    stop and tail lamp 96, 97
    switches 58, 96, 97, 99
    wiring 98
    wiring diagrams 114–119
**Electronic ignition system** 110, 111
**Engine:-**
    bearings 25
    crankcases 22, 33, 35
    crankshaft 22, 25, 107
    cylinder head and barrel 19, 27, 40
    dismantling – general 19
    examination and renovation – general 24
    fault diagnosis 42
    left-hand casing 20, 33, 41, 106
    oil:
        level check 7
        level gauge 109
        pump 9, 41, 52–54, 109
        tank 54
    piston and rings 19, 25, 39
    reassembly – general 33
    refitting in the frame 40, 41
    removal from the frame 15, 16
    right-hand casing 21, 33, 35, 106
    specifications 14, 102
    starting and running the rebuilt engine 42
**Exhaust** 9, 52, 109

## F

**Fault diagnosis:-**
    brakes 92
    clutch 43
    electrical 100
    engine 42
    frame and forks 75
    fuel and lubrication 55
    gearbox 43
    ignition 60
    wheels and tyres 92
**Filter – air** 9, 50, 108
**Final drive chain** 8, 90
**Flywheel generator:-**
    refitting 39
    removal 19
    testing 56, 94
**Footrests** 73
**Frame** 68
**Front brake:-**
    drum:
        adjustment 9, 84
        examination and renovation 78
        wear check 11
    hydraulic:
        bleeding 84
        caliper 80, 112
        examination and renovation 78
        fluid level check 7
        master cylinder 83
        pad renewal 80, 112
        pad wear 11
**Front forks:-**
    dismantling 63, 64, 111
    examination and renovation 65
    oil change 10
    reassembly and refitting 65, 111
    removal 61, 62
    specifications 61, 104
**Front wheel:-**
    bearings 11, 85
    check 10, 76
    removal 77
**Fuel system:-**
    air filter 9, 50, 108
    carburettor 44–48, 108
    fault diagnosis 55
    petrol tank and tap 45, 107
    reed valve 50
    specifications 44, 103
**Fuse** 95

## G

**Gearbox:-**
    examination and renovation 27
    fault diagnosis 43
    modifications 107
    oil change 10
    oil level check 7
    shafts:
        refitting 33
        removal 22
    specifications 15
**Gearchange mechanism:-**
    internal components:
        modifications 107
        refitting 33
        removal 22
    examination and renovation 27
    external components:
        refitting 35, 106
        removal 22, 106
**Generator** *see* **Flywheel generator**

## H

**Handlebar switches** 96
**Headlamp** 95, 113

# Index

**Horn** 98
**HT coil** 58, 110
**Hydraulic brake** see **Brakes**

## I

**Ignition system**:-
    CDI unit 108
    checking 56
    condenser 58
    contact breaker 9, 57, 59
    electronic ignition system 110, 111
    fault diagnosis 60
    HT coil 58, 110
    pulser coil 110
    source coil 110
    spark plug 60
    specifications 56
    switch 58, 99
    timing 9, 57, 111
**Instruments**:-
    bulbs 97, 113
    drives 73, 74, 106
    heads 73

## K

**Kickstart** 22, 35, 73, 106

## L

**Lamps** see **Bulbs**
**Legal check** 9
**Lubricants – recommended** 13
**Lubrication system**:-
    fault diagnosis 55
    oil pump 9, 41, 52–54, 109

## M

**Main bearings** 25
**Maintenance – routine** 7–11
**Master cylinder** 83

## O

**Oil – engine**:
    level check 7
    level gauge 109
    pump 9, 41, 52–54, 109
    tank 54
**Oil – front forks** 10
**Oil – transmission** 7–10

## P

**Parts – ordering** 6
**Pedal – rear brake** 73
**Petrol tank** 45
**Petrol tap** 45, 107
**Piston**:
    examination and renovation 25
    refitting 39
    removal 19
**Points – contact breaker** 9, 56

**Prop stand** 70
**Pulser coil – ignition** 110
**Pump – oil** 9, 41, 52–54, 109

## R

**Rear brake**:-
    adjustment 9, 88
    pedal 73
    wear check 11
**Rear chain** 8, 90
**Rear suspension units** 70
**Rear wheel**:-
    bearings 11, 85
    cush drive 88
    examination and renovation 86
    modifications 113
    removal and refitting 86
    sprocket 88
**Rectifier** 94
**Reed valve** 50
**Rev counter** see **Tachometer**
**Rings – piston** 25
**Routine maintenance** 7–11

## S

**Safety checks** 5, 9
**Seat** 73
**Source coil – ignition** 110
**Spark plug** 60
**Specifications**:-
    brakes 76, 104
    clutch 15, 103
    electrical 93, 104
    engine 14, 102
    frame and forks 61, 104
    fuel 44, 103
    gearbox 15, 103
    ignition 56, 103
    lubrication 44, 103
    wheels and tyres 76, 104
**Speedometer**:-
    bulbs 97, 113
    drive 73, 74
    head 73
**Sprockets** 88, 90
**Stands** 70
**Steering head** 62, 66, 111
**Suspension**:-
    front 10, 61–65, 111
    rear 68, 70, 111
**Swinging arm** 68, 111
**Switches**:-
    handlebar 96
    ignition 58, 99
    lighting 99
    neutral 20, 35
    stop lamp 97

## T

**Tachometer**:-
    bulbs 97, 113
    drive 73, 74, 106
    head 73
**Tail lamp** 96
**Tank**:-
    oil 54
    petrol 45

## Index

**Tap – petrol** 45, 107
**Timing – ignition** 9, 57, 111
**Tools** 12
**Torque wrench settings** 13, 84, 105
**Tyres:-**
   removal and refitting 91
   specifications 11, 76, 104

### V

**Valve – reed** 50

### W

**Wheel:-**
   alignment 90
   bearings 11, 85
   examination and renovation 10, 76, 86
   modifications 113
   removal and refitting 77, 86
**Wiring diagrams** 114–119
**Working conditions** 12

### Y

**YEIS unit** 109

Printed by
**J H Haynes & Co. Ltd**
Sparkford  Nr Yeovil
Somerset  BA22 7JJ  England